Kanban for the Supply Chain

Fundamental Practices for Manufacturing Management

Stephen C. Cimorelli, CFPIM

PRODUCTIVITY
productivity press

Productivity Press
NEW YORK, NEW YORK

Most Productivity Press books are available at quantity discounts when purchased in bulk. For more information contact our Customer Service Department (888-319-5852). Address all other inquires to:

Productivity Press
444 Park Avenue South, 7th floor
New York, NY 10016
United States of America
Telephone 212-686-5900
Fax: 212-686-5411
E-mail: info@productivitypress.com

Library of Congress Cataloging-in-Publication Data

Cimorelli, Stephen C.
   Kanban for the supply chain : fundamental practices for manufacturing
management / Stephen C. Cimorelli.
     p. cm.
   ISBN 1-56327-314-4 (alk. paper)
   1. Just-in-time systems. 2. Production management. 3. Business logistics. I. Title.
TS157.4.C56 2005
658.5'1--dc22

                                           2005023956

09 08 07 06 05 5 4 3 2 1

DEDICATION

I wish to dedicate this book to my wife, Cindi, and children, Stephen and Elizabeth. I am not a writer by trade, and therefore worked many nights and weekends over a period of more than two years, time which should rightly have been spent with them. Their understanding, support, and encouragement have made this book possible. They have enriched my life beyond measure, and I hope are proud of what I have accomplished. I would be remiss not to mention "Ranger the Dog," who has been my constant companion and most ardent fan, and who forced me up from my desk for walks and games when we both needed them. He has taught me well.

CONTENTS

ACKNOWLEDGMENTS

As with most worthwhile efforts, I could not have written this book alone. Beginning in the mid-1980s, when I first learned of JIT and kanban, I have sought opportunities to implement the principles and techniques they espouse. It was not until 1995, when I joined Square D - Schneider Electric, that I actually became part of an organization that embraced those principles and was working to master them. Over the past ten years, I have become a vocal advocate of kanban pull processes. I have also been presented with opportunities to educate others in kanban processes, concepts, and principles, and have been entrusted to help implement kanban techniques across the company. In the past four years, I have helped to do the same in nearly 20 manufacturing plants in the United States, Canada, and Mexico (both inside and outside of Square D), and in product lines ranging from small high-volume repetitive circuit breakers to low-volume make-to-order electrical distribution equipment large enough to fill a room. Kanban has a place in each of these!

In addition, I was blessed to be part of a team the likes of which I had never seen before; people who selflessly shared their knowledge and experiences with me. I wish, in particular, to acknowledge the following: Bob Taylor, for pounding the basics and theories into my head and for helping me to see clearly how they work in practice; Mark Harris, for clarifying the proper role of kanban and for explaining when to apply different techniques; Mike McCreery, for sharing his expertise in the systems that provide the data necessary for kanban analysis and which enable electronic kanban techniques; and finally, Terry Haire, for his leadership, technical and organizational insights, support and example. I must also thank these individuals collectively for their trust and friendship. This book is largely the result of their combined knowledge and expertise, and I am truly in their debt.

Finally, I would like to thank those who participated in shaping this material. Michael Sinocchi, acquisitions editor for Productivity, recognized the need for this workbook and encouraged me to focus it on supply chain management (SCM). Gary Peurasaari, Productivity's content editor, helped develop this material into a workbook and helped round out its SCM and lean manufacturing content.

INTRODUCTION

This workbook is for manufacturing supply chain management professionals *wanting to put kanban into practice to improve material flow.* It is for those who are ready to stop thinking about a conversion from materials requirements planning (MRP) push techniques to kanban pull techniques and want to make it happen now! In particular, this workbook focuses on purchased parts—pulling them into the operation and synchronizing them to actual consumption in the manufacturing process. (For those interested in implementing kanban through the manufacturing process itself, I recommend *Kanban for the Shopfloor* by Productivity Press.) Unlike MRP, which attempts to integrate planning and execution—pushing materials into the operation according to the plan, kanban de-couples these two activities. It gives supply chain managers the ability to *plan* inventory replenishment, then *pulls* inventory into the operation according to actual consumption. Supply chain planners in a kanban process use the same base of information as their MRP counterparts to establish appropriate kanban parameters for each part, i.e., *a plan for every part.* However, when these parameters are established, *kanban controls the execution,* triggering purchase orders to replenish *actual consumption.* Through an iterative process of planning, adjusting, and executing, kanban pull techniques keep inventory levels synchronized with demand and the supply chain synchronized with actual production.

It is not simply enough, however, to *synchronize,* because synchronizing to erratic demand patterns may not be feasible either for the internal operation or for suppliers. To be most effective, a company must *stabilize* the manufacturing operation. This requires a total supply chain view, which includes customers, internal operations, and suppliers. In Chapter 11, we will explore a specific technique for stabilizing production (and thereby the supply chain), along with a general discussion of leveling strategy.

A key success factor to implementing kanban is having clear, consistent communication, with all parties abiding by the same techniques and using the same terminology. Anyone experienced in supply chain management knows the pitfalls of trying to carry out a production plan without first communicating and coordinating the process with suppliers. Beginning with the *decision* to implement kanban and in every phase thereafter, supply chain managers play a pivotal role in communicating to suppliers what the company expects of them and the tools and techniques the suppliers will need to make the conversion to kanban successful. Throughout this workbook, as we explore kanban practices that affect the supply chain, I will provide specific communication opportunities to assist supply chain management (SCM) teams in their efforts.

The figures and commentary in this book are derived from real-life experiences implementing kanban in a variety of manufacturing environments. Other authors have written more extensively on kanban, many with a focus on production and assembly processes *inside of a plant*. *The Toyota Production System*[1] remains the seminal work, not only on kanban but also on a wealth of lean manufacturing and just-in-time processes. Although every SCM professional should be familiar with and have access to this book, my personal experience has been that while the Toyota Production System (TPS) is an excellent source of specific techniques and processes, it does not provide a clear, simple, roadmap for implementing kanban and the underlying processes that enable kanban to work effectively.

My goal for this workbook is to provide SCM teams with just such a roadmap, one that will facilitate putting fundamental kanban concepts to work in a way that is *immediately actionable and immediately enhances supply chain management*. To this end, I will first present some concepts that are critical to implementing kanban successfully, then demonstrate kanban's practical application, and finish with some key principles for sustaining the kanban system and improving its effectiveness through level-loading. Along the way, I will provide the SCM team with exercises, lean notes, and communication notes relating to supplier relations. My intent is to present kanban fundamentals in such a way that once grasped, kanban can immediately be applied by SCM teams to improve material flow, thereby increasing manufacturing productivity and the profitability of their companies.

Work Book Outline and Style

Figure I-1 provides a broad view of the content of this book. Chapters 1 and 2 focus on kanban and supply chain management foundational concepts and definitions. Chapter 3 reviews the Pareto 80/20 rule and challenges you to apply it to *your company's* parts through an ABC classification exercise that the SCM team will draw on for the remainder of the workbook. Chapter 4 presents a graphical model of material

---

1. Ohno, Taiichi, *Toyota Production System: Beyond Large-Scale Production*. Portland, OR: Productivity Press, 1988

replenishment patterns, depicted as a saw tooth diagram, which is at the heart of kanban. We will use this model extensively, especially in Chapter 7, to help the SCM team visualize what actually takes place on the shopfloor as material is purchased, consumed, and replenished in a recurring cycle. Team exercises, structured in a way that allows you to use your own data, will reinforce these principles.

- Overview of Kanban in a Lean Environment
- Supply Chain Management Fundamentals
- Applying the ABC Classification
- The Saw Tooth Diagram—Analyzing Inventory Behavior
- Lead Time / Lot Size Guidelines
- Statistics 101 and Demand Variability
- Saw Tooth Exercises
- Physical Techniques of Kanban Replenishment Systems
- Kanban Maintenance
- Kanban Implementation Approach
- Stabilizing Production

Figure I-1. Chapter Outline for Discussion of Kanban Fundamentals

Throughout this workbook, we will explore a variety of scenarios using the saw tooth model. Each scenario will lead the SCM team through a simple but powerful study of material flow cycles, while simultaneously analyzing some common challenges and providing alternative solutions for each. The SCM team will be challenged to examine similar scenarios in their own companies and discuss how the solutions may help them. Chapter 5 deals with guidelines for setting lead time and lot sizes using the Economic Order Quantity (EOQ) formula and other practical decision rules based on earlier ABC analysis. Chapter 6 introduces some basic statistics to determine how to analyze demand variability and use it to calculate safety stock.

Beginning with Chapter 8, the SCM team will move from the conceptual to the physical challenges of implementing kanban. These include using kanban's manual-visual control processes and applying them to various single-bin, multi-bin, and other binless techniques. Recommendations for how to apply each appropriately are provided.

Chapters 9 and 10 discuss how to establish roles and responsibilities to implement kanban effectively and provide specific approaches for maintaining and improving the kanban system over the long term. One of these long-term keys to success is level-loading, discussed in Chapter 11.

Finally, a word about style. This workbook uses both instructive and narrative materials. It provides SCM professionals leading the kanban transformation in their companies with a ready source of basic materials: 70+ figures and tables accompanied with explanations and examples will eliminate the need to create your own educational and training materials. Having presented this material in a variety of forums, including training classes, APICS workshops, and dinner meetings, I have attempted to capture some of the dialog that normally evolves during these sessions. My hope is that this workbook will aid a broader audience in understanding and implementing kanban pull processes to improve the supply chains, which are so critical to lean plant operations.

About the CD

Included with this book is a CD that contains exercises, "communication notes," and figures presented in a PowerPoint™ format. This material will greatly assist SCM team leaders conducting training sessions with team members in a classroom or workshop environment.

CHAPTER ONE

Overview of Kanban in a Lean Environment

The next two chapters provide a primer to help the SCM team prepare to implement kanban. Though the material is basic, it is important that everyone on the SCM team, as well as external suppliers, is on the same page, especially in using terminology and understanding how kanban fits in with just-in-time (JIT) and an overall lean environment. Of course, the place to start is by defining kanban.

The Japanese word "kanban" means card, ticket, sign, or signboard. Kanban originated in the Toyota Production System (Toyota's lean enterprise system), as a tool for managing the flow of production and materials in a JIT "pull" production process. JIT is one of the two pillars of lean. The other is jidoka, which means building in quality and designing operations and equipment so that people are not tied to machines and are free to perform value-added multitask work that is appropriate for humans. In a pull system, a process makes more parts only when the next process withdraws parts—in effect "pulling" the parts from the earlier process when needed. Jidoka supports the pull process by ensuring that only quality parts are passed to the next operation and stopping production when that is not the case, driving immediate corrective action. (You can explore these principles in Productivity Press' shopfloor series, *Just-in-Time for Operators and Kanban for the Shopfloor*).

There are two important aspects to kanban: production and material replenishment. The APICS Dictionary (11th ed., 2005) defines kanban as "a method of just-in-time production that uses standard containers or lot sizes with a single card attached to each." The objective of this *production aspect of kanban* is to balance workflow by signaling production of a part, component, or subassembly only when the next operation in the process has begun work on the unit or lot previously produced. In this way, material is *pulled* through the process in a synchronized fashion. The product's container often serves as the kanban signal in this model, with an attached card providing additional information if needed.

The second part of the APICS definition of kanban is "a pull system in which work centers signal with a card that they wish to withdraw parts from feeding operations or suppliers." This is the *material replenishment aspect of kanban*, in which material is pulled into a work center from internal or external suppliers. Suppliers are required to provide materials in the right quantity within a designated lead time. They need to do this reliably, consistently, and with the assurance of high quality. Advantages of kanban include:

- **Simplicity:** Kanban provides clear and precise manual-visual control processes.
- **Lower cost:** Production and move signals utilize low cost visual tools.
- **Agility:** Pull processes respond quickly to changes in customer demand.
- **Reduced inventory:** Kanban limits overcapacity in processes and avoids overproduction of inventory, driving less "just-in-case" inventory.
- **Minimal waste:** Kanban minimizes the wastes of overproduction, unnecessary inventory, and accompanying floor space.
- **Improved manufacturing productivity:** Kanban maintains control of the production line, synchronizing all steps in the process.
- **Delegation of responsibility:** Kanban's manual-visual control processes give responsibility to operators to make and act on production and inventory replenishment decisions.
- **Improved communication:** In the context of a JIT environment, kanban makes clear what managers and supervisors must do. It reduces "management by shortage lists" and expediting.
- **Accelerated improvement:** Kanban's underlying analysis processes promote and sustain continuous improvement.
- **Enabling JIT to operate:** Kanban provides two essential elements to JIT, the ability to manage flow and the ability to control inventory.

What Kanban Does

Kanban provides a means of pulling material into, and through, a manufacturing process. The material being pulled can include raw materials, parts, components, and subassemblies, including parts manufactured in house or purchased from outside suppliers. The major elements of a kanban system include the following:

- What to pull
- When to pull it
- How much to pull
- Where to pull (from/to)

Throughout this workbook, I will examine each of these elements in detail and discuss practical approaches to manage each within the system.

> **Lean Note:** In a pull system, a process makes more parts only when the next process withdraws those parts. Pull begins with the leveled production schedule for the final process, which is based on actual or expected customer orders. The final process uses kanban to pull needed parts from the previous process, which pulls from the process before it, and so on. Ideally, the entire line will be balanced to takt time. Takt time is the "drumbeat" of the line and represents the time allowed to produce one unit of product through each operation and ultimately off the end of the production line. Balancing to takt is part of an overall lean strategy, and improves the effectiveness of kanban. It is not a prerequisite to implementing kanban.

Kanban works best when applied to repetitively used material when future demand is predictable and expected to remain relatively stable. But it is not limited to these conditions. Factors such as cost, annual quantity, lead times, and lot sizes all play a role in determining if kanban is the right tool for a given part. (Chapter 6 examines this point in detail). Furthermore, a company needs to have some JIT tools and techniques in place to ensure a successful kanban conversion and to sustain it long term. I will touch on a few of these points at the end of this chapter.

Kanban is essentially an order point system whereby the user (internal or external customer) of a particular material (part, component, subassembly, etc.) signals to the supplier when more material is needed. The signal is called the kanban and can take many forms, but it is generally some type of card. The most common is a small form inserted in a rectangular vinyl envelope. The form lists, in bar code and in text, pertinent information, such as how many of what part to pick up or which parts to assemble. A typical kanban card provides:

- **Supply information:** Supplier name and number (vendor code) and stocking location
- **Part information:** ID number, description, quantity
- **Customer information:** User group and location, storage locations, kanban number

The kanban card may also provide:

- Raw material
- Pickup information
- Transfer information
- Other information deemed important by the company or process owners

SCM professionals use the kanban to move the information vertically and horizontally within the company and between the company and external suppliers. There are three types of kanban cards, each of which contain information relevant to its purpose. Since this workbook focuses on managing the supply chain with kanban, the sample cards shown below represent typical external supplier and move kanbans (see Figures 1-1 and 1-2).

1. *Move kanban:* Authorizes a process (workstation or cell) to retrieve parts from the previous process that is responsible for producing the item. (Key information: Part Number, Part Description, Quantity, From Location, To Location)

2. *Production kanban:* Authorizes the previous process (workstation or cell) responsible for replenishment to produce the item. (Key Information: Part Number to be produced, Part Description, Quantity to produce, Stocking Location of the finished product, Process Location and description)

3. *Supplier kanban:* Authorizes an outside supplier to deliver more parts to the process (workstation or cell) that is using supplier components. (Key Information: Part Number, Description, Supplier, Quantity, and Stocking Location once parts are received)

Companies typically convert external supplier kanbans to purchase orders, using their normal procurement ordering systems. When the parts arrive, they are matched to the kanban card and put in stock or delivered to the line. (If parts are put in stock, the production line will signal the need for parts to be brought from stock via a move kanban). Similarly, internal suppliers typically release some type of manufacturing order (enabling other internal control and measurement processes) and match finished parts/subassemblies to the kanban card for delivery to the ordering group.

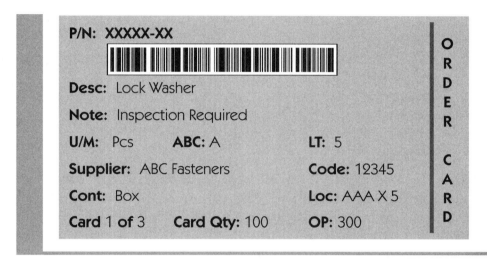

Figure 1-1. Example of External Supplier Order Kanban Card

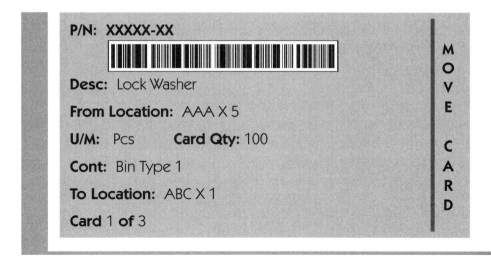

Figure 1-2. Example of Move Kanban Card

The order point at which the user sends the kanban must be sufficient to cover "demand through lead time," which is the anticipated demand for the material during the time required for the supplier to replenish it. If the kanban signal is between two operators on an assembly line, the order point will be quite small, perhaps even zero, if the supplying operator can replenish the material before the downstream operator needs the next unit. This is the classic case where the kanban signal is an empty square painted on a work surface or on the floor between two operations. As long as parts are in the square, no further production is allowed by the producing work center. Only when the square is empty is the producer authorized to make more material and replenish the empty square.

If the signal is between the company and an external supplier, and the lead time to replenish the material is several days, weeks, or longer, the operator must send the kanban signal early enough to allow the supplier to replenish before a stock out occurs. (A stock out occurs when there is no material on hand to cover an immediate need; it is sometimes referred to as a shortage). Therefore, the order point must be high enough to cover expected demand during the replenishment lead time; additionally, there must be sufficient safety stock to cover *variability* of demand during the lead time. For this reason, companies need to include external suppliers in the planning phase, so they can deal effectively with changes to lead time, lot sizing, stocking plans, transportation, forecasting, order processing, or similar issues. Suppliers, of course, should participate in the decision of how kanban orders will be signaled. A well-designed kanban signal will include location information indicating who and where the supplier is as well as who and where the user is, so that those responsible for transferring the signal and the material know exactly where both parties are and precisely where the material is stored.

> **SCM Communication Note:** Contact all external suppliers to inform them about the conversion to kanban, what impact it may have or may not have on them, as well as your company's expectations. Teaming with suppliers early in the process and maintaining regular communication afterwards are two keys to success. Furthermore, companies can provide kanban training to their suppliers, further improving the performance of the total supply chain.

From the descriptions above, it should be clear that kanban is very much a manual-visual control process. In this regard, you can think of it as part of a holistic visual factory system using 5S, andon, color-coding, poka-yoke, etc., that quickly communicates information that everyone can understand. However, viewing kanban as simply one of these visual lean tools would be to look at it only skin-deep. Beneath the surface of kanban's replenishment techniques lie all the elements of a successful SCM system that can create, manage, and assure the flow and production of materials, including:

- Demand planning and forecasting
- Production planning
- Analysis of kanban parameters
- Software systems for order management
- Bills of material
- Finance and accounting
- Sound supplier management

This workbook will not examine all of these in detail; however, each is important, and I will discuss relevant linkages along the way.

Finally, to be successful, it is important for the SCM team to understand and continuously apply the seven basic rules of kanban, which are listed below.

1. *The later process goes to the previous process to withdraw only what it needs.* Where travel distances are likely to disrupt production, dedicated material handlers will transport kanban cards and parts, enabling production workers to focus on value-added activities.

2. *The previous process makes only the quantity needed to replace what the later process removed or rounds the production quantity to an agreed-upon standard lot size.* The kanban card indicates this quantity.

3. *Never send defects to the next process.* This means operators (or those responsible) must stop the production to correct any problems causing the defects. In a low-inventory production system, this motivates operators to implement preventive improvements, such as mistake proofing (poka-yokes) and autonomation (jidoka).

4. *A kanban must always accompany parts and products on the line.* Since only a certain number of kanbans exist, they serve as a visual control of the amount of inventory allowed in the work area.

5. *Equalize production.* Production quantities must be leveled (load-smoothing) to avoid fluctuation and eliminate waste. Spreading the

quantity evenly over time and balancing lines to takt time help assure a smooth flow between processes.

6. *Use kanban to fine-tune the schedule.* Since production occurs when instructed by the kanban, small increases or decreases in the amount(s) to be produced can be handled easily by changing how often kanbans are transferred between processes. Kanban systems are not well suited to handle large fluctuations in customer demand. One exception to this is low-cost parts, either purchased externally or manufactured in house. These parts can be held in a storeroom until they are pulled into production by a move kanban. Relatively high levels of safety stock are required, but since the cost is low, little inventory investment is required, and this investment may offset the costs of other replenishment processes.

7. *Stabilize, rationalize, simplify, and standardize the process.* Like rule 3, this rule is about improvement—in this case, improving the process to avoid waste and unpredictability.

Definitions—Building a Common Language

To develop a common terminology within the company and with outside suppliers, it is essential to define the key terms (see Figures 1-3 to 1-8). For now, the definitions of these terms simply stand alone, but as the SCM team participates in the practical applications of kanban, it will become clear how these terms are essential to understanding, communicating, and applying kanban fundamentals.

Kanban

- Japanese word meaning card or sign.

- A method of just-in-time production that uses standard containers or lot sizes with a single card attached to each.

- A pull system in which work centers signal with a card that they wish to withdraw parts from feeding operations or suppliers.

Figure 1-3. Kanban Definition from APICS Dictionary-9th Edition

Building a Lean Environment

A few comments are in order regarding JIT, takt, and backflushing. Stepping back from kanban fundamentals and looking at the broader picture,

Just-In Time (JIT)

- JIT is a set of principles, tools, and techniques that helps to produce a variety of products in smaller quantities, with a shorter lead time, to meet specific customer needs.

- The key goals of JIT and lean are the same; defining a **value stream, creating flow, pull**, and promoting **continuous improvement**, but JIT usually refers to the system of producing and delivering the right items at the right time in the right amounts.

- JIT uses the kanban system to manage the pull production system and help coordinate the production and movement of parts and components between processes to avoid excesses or shortages.

- JIT requires confidence that the supplier chain will meet commitments.

Figure 1-4. Just-in-Time Definition

Inventory Management

- **Order Point (OP).** An established inventory level that when reached, signals the need to issue a replenishment order.

- **Safety Stock (SS).** A quantity of inventory planned to be on hand to protect against fluctuations in demand.

- **Lead Time (LT).** The time required to replenish inventory. This is normally measured as the number of days from when the order point is broken, to receipt and put-away of the corresponding replenishment order. May include order processing time, supplier lead time, and receipt and put-away time.

Figure 1-5. OP, SS, and LT Definitions

it is important to understand that kanban works within the framework of JIT to manage flow. This means that a company cannot achieve JIT without some form of kanban. Conversely, if a company is not pursuing JIT or a lean environment, kanban is not likely to succeed. We will look at a few points regarding the lean environment in the next section.

Inventory Management

- **Lot Size (LS).** An established order quantity representing an agreed-upon amount of inventory in a replenishment order. (LS is used in conjunction with a unit of measure, such as pieces, pounds, rolls, square feet, etc.)

- **Demand through Lead Time (DLT).** The amount of inventory expected to be used during the replenishment lead time. DLT is normally based upon an average usage rate over a predetermined period of time (e.g., average monthly wage).

- **ABC Analysis.** A method of classifying inventory by Pareto Analysis, the 80/20 rule based on the premise that 80% of value is held by 20% of the population. (Pareto analysis is sometimes referred to as the 80/20 rule, ABC analysis, and ABC classification.)

Figure 1-6. LS, DLT, and ABC Analysis Definitions

Inventory Management

- **Bill of Material (BOM).** A listing of all raw materials, parts, intermediates, and subassemblies required to make a parent assembly, showing the quantity of each time.

- **Backflush.** The deduction from inventory records of the components used in an assembly or subassembly by exploding the BOM by the production count of assemblies produced. (A post-deduct inventory transaction process).

Figure 1-7. Bill of Material and Backflush Definitions

Production Management

- Takt is a German word meaning either clock or musical meter. Takt time (TT) is the rate at which each product needs to be completed to meet customer requirements and is expressed in seconds (or other time unit) per part. It is the beat or pulse at which each item leaves the process.

- TT is the tool that links production to the customer by matching the pace of production to the pace of actual final sales.

- TT is determined by taking the available production time and dividing it by the rate of customer demand. In other words, it is the daily operating time divided by required quantity per day (unit).

Figure 1-8. Takt Time Definition

Takt Time

Figure 1-9 provides an example of a balanced assembly line in which each workstation completes its part of the process within takt time and products flow off the end of the line at the same pace.

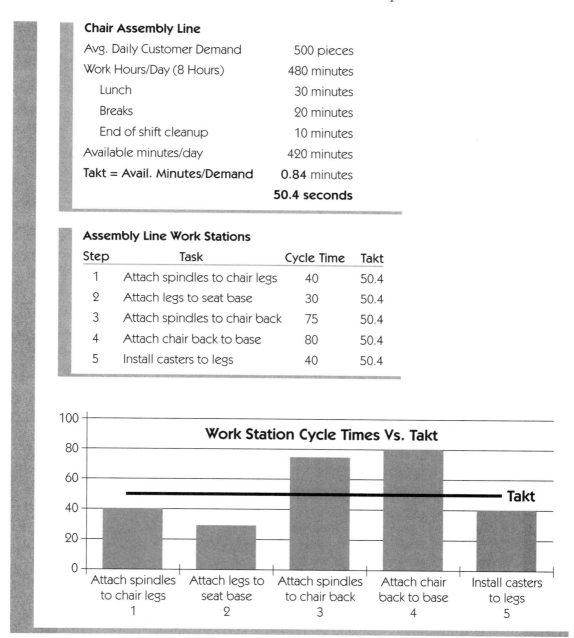

Chair Assembly Line

Avg. Daily Customer Demand	500 pieces
Work Hours/Day (8 Hours)	480 minutes
Lunch	30 minutes
Breaks	20 minutes
End of shift cleanup	10 minutes
Available minutes/day	420 minutes
Takt = Avail. Minutes/Demand	**0.84** minutes
	50.4 seconds

Assembly Line Work Stations

Step	Task	Cycle Time	Takt
1	Attach spindles to chair legs	40	50.4
2	Attach legs to seat base	30	50.4
3	Attach spindles to chair back	75	50.4
4	Attach chair back to base	80	50.4
5	Install casters to legs	40	50.4

Figure 1-9. Example of Takt Time for Establishing a Balanced Line

The calculation of takt time is simply the available work time on the assembly line divided by the average daily demand from the customer. In the example shown in Figure 1-9, takt time is 50 seconds, indicating that each workstation on the assembly line must be able to complete its work content in 50 seconds or less. As the graph indicates, workstations 3 and 4 exceed takt, making it impossible for the line to meet customer demand. Two things must be done to balance the line: 1) eliminating non-value added time (waste) from these tasks and 2) moving some of the work content to workstations that are below takt. In addition, workstations whose work content remains well below takt represent opportunities to remove cost from the assembly process. If some of these operations can be combined and still remain below takt, you can accomplish the assembly process with fewer operators.[1]

Once the line is balanced, you can use kanban to pull product through the line at a pace consistent with customer demand. Doing so also provides pull signals back upstream to suppliers, with these signals being balanced to customer demand. To improve the overall process, the production schedule of the line should be level-loaded to average customer demand. Chapter 11 presents a technique for level-loading, which, when combined with balancing the line to takt and implementing kanban, balances and level-loads the entire supply chain.

One final comment on takt time deserves mention. Balancing assembly lines to takt is not a prerequisite to using kanban. Implementing kanban to pull materials into the assembly line is still possible and should be pursued. However, shopfloor kanban and supplier kanban results will be much better if kanban implementation and line balancing occur simultaneously. Level-loading further improves the entire process. Do not allow yourself to be stuck in a cycle of paralysis, believing that any of the three techniques is a prerequisite to the others. Start somewhere! As you implement and refine these three elements of a lean enterprise, the synergies between the techniques will resonate throughout your supply chain.

Backflushing

Backflushing is also known as post-deduct processing. In a typical MRP environment, inventory is issued to orders, with discreet orders created to produce or purchase the part numbers at each level of the Bill of Material (BOM). Figure 1-10 provides a simple example of a BOM for finished part X. Let's assume there is an order for 50 Ys. This would require 100 Ws (note that the BOM requires 2 Ws for each Y) and 50 Vs

---

1. A detail discussion of takt time and line balancing is beyond the scope of this book. For further details, I recommend *Kanban for the Shopfloor*, and *Standard Work for the Shopfloor*, both by Productivity Press.

to be issued from stock to the order. In an MRP environment, this usually involves an inventory transaction, which deducts the quantities issued from each part's on-hand balance. When the order is completed, the finished quantity of Ys is placed in inventory. Again, an inventory transaction is required to increase the on-hand quantity of Ys. Whenever another order for X is processed, this process is repeated. In this case, Ys and Zs will be issued to the order, decreasing the on-hand quantity of each, and the finished product X will either be sold or placed in finished goods inventory.

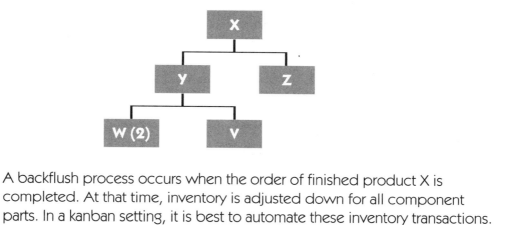

A backflush process occurs when the order of finished product X is completed. At that time, inventory is adjusted down for all component parts. In a kanban setting, it is best to automate these inventory transactions.

Figure 1-10. Bill of Material (BOM) for Finished Product X

In a kanban setting, it is useful to automate these inventory transactions. Kanban relies on manual-visual inventory maintenance and therefore does not require real-time updates to on-hand inventory balances. The transactions required to maintain this information can be quite time consuming, and the time saved in eliminating them can be put to better use.

The backflush process occurs when the order of finished product X is completed. At that time, the BOM is exploded, driving the inventory transactions for all components. For example, if 50 Xs were completed, then 50 Ys, 50 Zs, 100 Ws, and 50 Vs must have been consumed, so the backflush process will automatically perform the inventory transactions to deduct these balances.

JIT Preparation for Kanban

By producing only what is needed when it is needed, kanban eliminates buffer inventories and communicates customer demand upstream to trigger process steps just-in-time to meet demand. By acquiring products at the time needed in the quantity needed, a company eliminates waste and improves efficiency. As stated earlier, kanban is part of a holistic, lean environment, in which JIT practices operate in tandem with other lean principles. The kanban system's main role is to manage flow and control inventory in a JIT environment. If a company is not actively improving the corresponding manufacturing processes, kanban will most likely fall short of its potential benefits and may even fail. For optimal kanban and improved supplier relations, a company must eliminate some of its own non-value-added activities and operate in (or be moving in the direction of) a lean environment. This would mean eliminating waste in processes such as excess WIP (inventory), walking, conveyance, downtime, lead times, and defects.

The continuous activity of eliminating *muda*, or waste, is one of the foundations of lean manufacturing. Simply put, anything that does not add value to a product or process is waste. Figure 1-11 outlines what lean practitioners commonly refer to as the "seven deadly wastes." I have provided a brief statement of how each "waste" relates to kanban.

Seven Deadly Wastes	Kanban Application:
1. Overproduction	Produce only what's needed when your customer needs it. (Kanban signal vs. MPS Forecast)
2. Inventory	
3. Waiting	Minimize Lead Times
4. Transportation	Parts Close to Point of Use
5. Processing	Minimize Repackaging/Kitting, unnecessary material handling
6. Motion	Ergonomic Storage Design
7. Defects	Demand Defect-free Parts

Figure 1-11. Seven Deadly Wastes and Kanban Application

- **Overproduction and Inventory.** The waste of overproduction deals directly with the kanban/JIT principle of producing only what your immediate internal or external customer needs. Any overproduction requires space, material handling, and storage that otherwise would not be needed and delays production of whatever your immediate customer *needs*. Overproduction leads directly to the waste of unneeded inventory but is not the only cause. Traditional MRP push-scheduling techniques, which produce products to a forecast, generate inventory that may never be sold or inventory that may need to be held for many days, weeks, or months. Inventory not immediately needed by a customer is waste.

- **Waiting.** Any activity workers engage in that does not add value to the product or process is waste. Times spent waiting for materials or for an upstream process in an unbalanced production line are examples of this waste. For this reason, you must time kanban pull signals to ensure that a product from the supplying operation arrives when it is needed by the ordering process. Ideally, kanban materials will arrive just in time, but a bit early is better than late. Keep in mind though, that routinely delivering early creates unneeded inventory, which is itself a waste. This principle is clearly demonstrated in Chapter 7.

- **Transportation.** Since transportation does not contribute anything to the value of the product or process, it is waste. Although some transportation is unavoidable, you must find ways to minimize it. For kanban, you will ideally minimize transportation to the point where you employ visual signals between internal customers and suppliers, enabling kanbans of one, in a single-piece flow line. One-piece flow arranges production in a continuous flow of products through the manufacturing process, one unit at a time, at a rate determined by the customer's need (takt time) with the least amount of delay and waiting. You can also minimize transportation by stocking externally supplied or internally batch-produced material, at or near the point of use.

- **Wasted motion.** This includes any wasted motion such as reaching or looking for parts, stacking parts, picking things up and putting them back down, carrying things, and searching for information. Ergonomic workstations and material storage devices can contribute greatly to minimizing or eliminating these wastes. Companies should identify and standardize best practices across the operation, not only in manufacturing processes but in material handling processes as well.

- **Defects.** Most defects have five direct causes: 1) inappropriate or poorly specified procedures or standards, 2) excessive variability in machines or steps during actual operations, 3) damaged or excessively variable materials, 4) worn machine parts, and 5) human error. Moving defective products through the process feeds all the other wastes. It causes overproduction to replace poor quality products and creates an inventory of unusable products. It also potentially creates the need for buffer replenishment inventory (extra safety stock). In addition, it duplicates waiting for the replacement "good" product, the unnecessary transportation of the defect, and all processing and motion associated with producing the defect. The goal is to attain zero-defects at all times.

No company can eliminate these wastes easily. Waste can, however, be minimized by implementing lean tools and principles and by focusing constant attention and dedication across the organization to sustain the effort. Once begun, the process requires a spirit of continuous improvement, or *kaizen*, to ensure that workers are practicing lean principles and that these efforts bring out the full potential of the operation and of the people working within that operation.

Begin with 5S

5S is a series of activities for identifying problems and eliminating wastes that contribute to errors, defects, and injuries. The SCM team can immediately implement these activities without having to wait for a company-wide lean initiative. Many lean practitioners view 5S as a foundational building block for creating a lean manufacturing environment. Figure 1-12 outlines the 5Ss, and briefly describes the linkages to kanban.

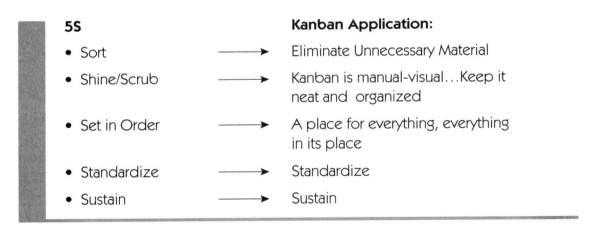

5S	Kanban Application:
• Sort	Eliminate Unnecessary Material
• Shine/Scrub	Kanban is manual-visual...Keep it neat and organized
• Set in Order	A place for everything, everything in its place
• Standardize	Standardize
• Sustain	Sustain

Figure 1-12. 5S and Kanban Application

The first S "sorts" out all unnecessary items from a work area, including tools, materials, documentation, etc. Remove anything that is not immediately needed to add value to the product or process. Next, is "scrub" or "shine," which deals with cleanliness and housekeeping. A clean, clutter-free work area contributes to a lean environment and to defect-free production, allowing you to detect and "scrub" things such as fluid leaks on equipment or dropped parts or tools immediately and keep materials clean and free of contaminants. This will also greatly reduce the likelihood of slips or falls. In general, a clean work environment tends to be a safe and productive one. The third S is "set in order," which means organizing all remaining items so they are readily available. Creating shadowboards, ergonomically placed materials, and visual aids are common activities in this step. "A place for everything and everything in its place" ensures that kanban materials are accessible with no wasted motion or searching.

The final two Ss have to do with *standardizing* and *sustaining* the improvements of the first three over time, throughout the operation. For kanban, standardizing how you store and replenish parts and selecting appropriate kanban signals to use simplifies the flow of materials throughout the factory. This in turn supports flexibility in work assignments and minimizes any confusion that may result in a less standardized environment.

Other important strategies a company can consider as a means to achieve a lean environment include JIT education, total productive maintenance (TPM), a quick changeover program, zero defects, cellular manufacturing, and a supplier performance rating. The SCM team can participate in many of these, and each should be part of a holistic JIT effort. They all work together synergistically to improve the entire lean operation.

> **SCM Team Exercise:** Take a moment to discuss the kanban terms in this chapter to establish a common language among team members and to determine what level of understanding each member has of kanban. Discuss both the general or broader meaning of these terms and how they may especially apply to your company and to JIT and lean thinking.

This chapter lays the groundwork for kanban fundamentals. The next chapter discusses a few fundamentals of supply chain management and how these fundamentals intersect with kanban.

CHAPTER TWO

Supply Chain Management Principles

As discussed in Chapter 1, it is important for the SCM team to understand the basic terms associated with kanban as well as the goals of implementing kanban. This chapter will outline some SCM principles used in conjunction with kanban to help the team operate effectively across the company and with outside suppliers. One of the keys to a successful implementation is for all parties to understand their respective roles and responsibilities. For the SCM team, this means having a clear understanding of the principles and role of supply chain management, as well as effectively communicating them across the company. For suppliers, it means understanding what a company expects of them. Optimal understanding is best achieved through clear communication and direction from the SCM team. This chapter highlights areas of communication that facilitate healthy working relationships with outside suppliers.

Defining the Supply Chain

The supply chain includes all processes from obtaining raw materials to final consumption or application of finished product by the end customers. All parties are linked through the flow of information, services, and products through three broad processes:

- Supply (raw materials and component parts)—how, where, and when you procure these items and supply them to manufacturing
- Manufacturing—how you convert raw materials and parts into finished products

19

- Distribution—how to use a network of distributors, warehouses, and retailers to ensure the finished products reach the final customers

Though this definition applies to all companies in the supply chain, not all supply chain managers necessarily face the same challenges and issues, because supply chains vary greatly from industry to industry and company to company. Having said this, there are several ways to cooperate with suppliers to use kanban effectively to synchronize and pull purchased parts into the operation where they are eventually converted into finished goods. These common denominators include supplier forecasting, lead time and lot size planning, and supplier performance measurement, as discussed below.

What is Supply Chain Management (SCM)?

Peter J. Metz defines supply chain management as:

> "... a process-oriented, integrated approach to procuring, producing, and delivering end-products and services to customers. It includes sub-suppliers, suppliers, internal operations, trade customers, retail customers, and end-users. It covers the management of materials, information, and funds flows."[1]

In any industry, SCM involves the careful planning and effective execution of activities and processes to enable a company to make informed strategic and tactical decisions along its entire supply chain, from raw material to distribution. Sound SCM practices can reduce costs, increase revenue, and improve quality, product design, and asset management. Inventory is one of a manufacturing organization's largest investments. For this reason, SCM practices have a significant impact on a company's return on assets (ROA), a key financial and operational measure. By pulling material through the operation in a manner that is synchronized to actual customer demand, kanban becomes a key component of the SCM process, minimizing inventory investment and thereby improving the overall asset management of the company.

SCM has a broad range of strategic and tactical applications, but for most purposes it utilizes the following components, each of which has some linkage to kanban as described below:

---

1. "Demystifying Supply Chain Management" Supply Chain Management Review, Winter, 1998, http://rac.alionscience.com/pdf/RAC-1ST/ISCM(1st).pdf

- **Demand Planning/Forecasting:** The SCM team must keep the kanban parameters, such as order points, safety stock levels and number of cards, in sync with forecasted demand. Techniques for doing so are discussed in detail later in the book.

- **Demand collaboration:** This entails a collaborative resolution process to develop consensus-based forecasts. Regular communications between marketing and operations, typically through a monthly sales and operations planning meeting, keep all functions' expectations of demand in sync, further improving the demand planning and kanban maintenance processes. In addition to demand forecasts, manufacturers should develop supplier forecasts and share them with suppliers on a regular basis.

- **Order Promising:** This occurs when a company can promise a product to a customer, taking into account lead times and constraints. By synchronizing the supply chain with actual customer orders, kanban improves the ability of manufacturers to meet order commitments.

- **Strategic network optimization:** The company must determine which plants and distribution centers (DCs) should serve which markets for each product. An annual or quarterly review of this strategy is typical. DC replenishment strategy is a key component of the collaborative demand planning process discussed above. Once distribution centers are established, applying pull replenishment processes to their stocked products further helps to synchronize the entire supply chain with real customer demand. Level-loading DC replenishment, especially on the highest-volume products, helps to balance demand signals on the manufacturing plants and smoothes the entire supply chain (see Chapter 11).

- **Production and distribution planning:** The company coordinates the actual production and distribution plans for the whole enterprise on a weekly or monthly basis. SCM teams can test short-term production plans against available capacity and inventory, assuring that kanban and DC replenishment signals will be achievable. They will be able to identify exceptions and develop work-around plans. SCM teams can use ERP systems, including material requirements planning (MRP), very effectively for planning purposes down to this level of detail.

- **Production scheduling:** The company creates a feasible production schedule on a daily or even a minute-by-minute basis for a single location. At this level of scheduling and execution, it is wise to de-couple MRP from the execution process, using kanban as the methodology to pull production through the manufacturing process and upstream through the supply chain.

- **Plan for the reduction of costs and manage performance:** This includes the process of planning, evaluation, accounting reporting, and quality reporting. SCM teams can use kanban fundamentals to identify inventory cost reduction opportunities. For example, they can set lead-time and lot-size objectives, allowing exception reporting of part numbers outside of accepted ranges, and calculate the corresponding inventory reduction opportunities. These points are examined in detail later in the book. Additionally, the team should have in place key performance indicators, such as supplier service and quality, stock-out tracking, and expedite tracking. The team should also establish goals for each and manage exceptions aggressively.

> **SCM Communication Note:** It is critical for supply chain managers to integrate all the partners in the supply chain by setting up the necessary information systems, such as Electronic Data Interchange (EDI) systems or web-based "real-time" tools and portals. Proper integration will give the company and its supplier network the necessary coordination for forecasting, demand balancing, inventory replenishment, order entry and tracking, and account management.

Traditionally, manufacturing, distribution, and purchasing organizations operated in silos. As companies depend more and more on external suppliers to deliver low cost, short lead times, and high quality, while simultaneously demanding greater flexibility, they will need to implement effective levels of planning and communications to meet and exceed customers' expectations. This requires supply chain managers to change the traditional combative mindset of buyer-supplier relationships to one of developing partnerships, which include clear communication, shared objectives, and efficient distribution channels.

In applying kanban pull techniques, this means establishing a cost effective and open information system with real-time communication and information flow to enable the supply chain to meet demand flexibility requirements while keeping inventory costs down. This is especially crucial in a "pull" system because purchase orders are based on actual, rather than expected, customer demand. (Remember that JIT builds to customer demand only.) As mentioned in the Introduction, applying kanban pull techniques will improve the supply chain system through an iterative process of planning, adjusting, and executing that will keep inventory levels synchronized to customer demand and keep the supply chain synchronized to actual production. In this way, kanban fundamentals become powerful principles and tools in the arsenal of supply chain management—something which all manufacturing supply chain managers can benefit from.

It is clear that kanban can help companies synchronize their suppliers to customer demand, but as mentioned in Chapter 1, for kanban to succeed, a company must also be pursuing other lean practices. If the company's house is not in good order, the SCM team's ability to communicate and partner with its suppliers will be fraught with many difficulties.

Lean Note: The heart of good supply chain management is for the company to work with suppliers on shared business goals built on mutual trust and respect, while challenging, training, and encouraging suppliers to use the necessary kanban and lean tools. A good supplier relationship is characterized by cooperation, stability, and mutually beneficial business practices.

In the next chapter, we will discuss the importance of the ABC classification, the method for classifying inventory by Pareto Analysis (the 80/20 rule). The SCM team will begin developing an ABC classification for its own company and will use the results for many of the team exercises in this workbook.

CHAPTER THREE

Applying the ABC Classification

Now that the SCM team has reviewed and discussed the fundamentals of kanban and SCM, and their role in a lean or JIT environment, it is time to look at some tools that can help put kanban into practice. The team's first step is to develop a clear picture of the parts inventory, then determine how each part affects the annual expenditure of the company. A useful inventory management technique is *ABC classification*, which uses the Pareto principle to classify parts based on the total annual expenditure for each. ABC classification determines which parts have the highest to lowest annual dollar volume and assigns an *A*, *B*, or *C* code to each part.

A-items represent the relatively small number of parts which have the highest annual dollar volume and therefore receive more attention than other parts. At the opposite end of the spectrum are the *C*-items, those low-value or slow-moving parts that make up the majority of part numbers. *B*-items are in the middle. Although *Bs* are less critical than *As* in terms of impact on inventory investment, and are fewer in number than *Cs*, they nevertheless represent a considerable inventory investment.

> **SCM Team Note:** The team can use the ABC classification to guide various inventory management decisions. These include how much safety stock to carry, how frequently to order, which kanban techniques (e.g., single-card or multi-card) to employ, and to determine if you should use kanban on a particular part number. We explore these decisions and techniques for each in detail later in the workbook.

A-items are sometimes referred to as "the critical few" and *C*-items as "the trivial many." However, I must include a word of caution on pushing these references too far. Anyone familiar with manufacturing knows that a missing *C*-item, no matter how "trivial," can shut down a production line as easily as a missing *A*- or *B*-item. Since *C*-items account for a small percentage of inventory investment, you can add additional safety stock to these parts with very little cost, ensuring their availability at all times and allowing more attention and control to be paid to higher-valued *A*- and *B*-items. Figures 3-1 and 3-2 show how the Pareto 80/20 rule (80 percent of the value in any population is held by 20 percent of the items) generally holds true for inventory. While *A*-items represent the 80/20 stratum, the bottom 50 percent of items at the other extreme account for only 5 percent of the total population's value, leaving the middle 30 percent of items valued at 15 percent.

Applying the 80/20 rule (80% of value in 20% of the parts) yields the following typical breakdown in an ABC classification:

ABC	%Items	% Cost
A	20	80
B	30	15
C	50	5
	100%	100%

Figure 3-1. ABC Classification—The 80/20 Rule

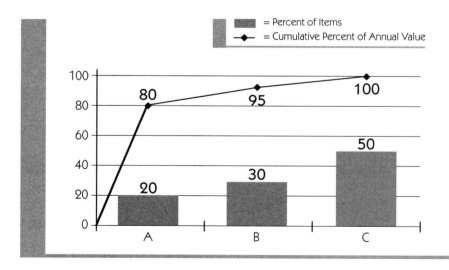

Figure 3-2. ABC Classification—Graphical Portrayal

Applying ABC Classification Step by Step

Tables 3-1 and 3-2 provide a step-by-step calculation technique for determining ABC classification. Table 3-1 shows that ZXY Company has 26 part numbers, ranging in value from 55 cents to $125, with annual usage quantities ranging from 33 to 6,410. It is critical to base each part's cost

Table 3-1. Company ZXY Example—Arranged by Part Number

- 26 Part Numbers
- Annual Quantity Usage from 33 to 6,410
- Costs from $0.55 to $125

Part Number	Annual Quantity	Cost	Annual Cost
A	249	$53.20	$13,246.80
B	409	5.44	2,223.31
C	204	3.40	693.60
D	352	1.50	528.00
E	270	2.30	621.00
F	6,410	1.70	10,897.00
G	264	3.40	897.60
H	57	7.44	424.02
I	77	6.70	515.90
J	428	53.75	23,005.00
K	247	0.69	170.49
L	337	5.50	1,853.50
M	282	2.30	648.60
N	535	1.12	599.20
O	573	6.00	3,438.00
P	35	6.18	216.44
Q	270	55.05	14,862.84
R	4,034	0.55	2,218.70
S	450	4.56	2,052.00
T	529	5.44	2,876.44
U	750	0.58	435.00
V	54	5.87	316.86
W	145	3.51	508.84
X	33	125.00	4,125.00
Y	2,010	1.25	2,512.50
Z	**1,1819**	**20.16**	**36,668.21**

Table 3-2. Company ZXY Example—Arranged by Annual Cost

ABC Process
- Sort by Descending Annual Cost
- Calculate Cumulative Cost & Cumulative Cost %
- Set ABC Break Points
- Calculate % Count Break Points as Sanity Check

Part Number	Annual Cost	Cum Cost	Cum %	ABC	% Cnt
Z	$36,668.21	$36,668.21	29%	A	
J	23,005.00	59,673.21	47%	A	
Q	14,862.84	74,536.05	59%	A	
A	13,246.80	87,782.85	69%	A	
F	10,897.00	98,679.85	78%	A	19%
X	4,125.00	102,804.85	81%	B	
O	3,438.00	106,242.85	84%	B	
T	2,876.44	109,119.29	86%	B	
Y	2,512.50	111,631.79	88%	B	
B	2,223.31	113,855.10	90%	B	
R	2,218.70	116,073.80	92%	B	
S	2,052.00	118,125.80	93%	B	
L	1,853.50	119,979.30	95%	B	31%
G	897.60	120,876.90	96%	C	
C	693.60	121,570.50	96%	C	
M	648.60	122,219.10	97%	C	
E	621.00	122,840.10	97%	C	
N	599.20	123,439.30	98%	C	
D	528.00	123,967.30	98%	C	
I	515.90	124,483.20	98%	C	
W	508.84	124,992.03	99%	C	
U	435.00	125,427.03	99%	C	
H	424.05	125,851.05	99%	C	
V	316.86	126,167.91	100%	C	
P	216.44	126,384.36	100%	C	
K	170.49	126,554.85	100%	C	50%

and usage figures on the same unit of measure. For example, if the part's cost is per piece, usage must also be in pieces; if cost is per pound, usage must also be in pounds. The first step in ABC analysis is to determine the annualized value of each part by multiplying the unit value by the annual usage. The key to the analysis is the annual value.

In the ZXY example, item K has the lowest annualized cost. Notice that it has neither the lowest unit cost nor the smallest annual usage. Neither of these is relevant on its own. What matters is the product of these two values (annualized cost).

Similarly, item Z has neither the highest unit cost nor the highest annual usage. It has, however, the highest annualized cost, making it the most important item in our ABC breakdown. A small change in inventory level on this part has a dramatic impact on plant inventory performance!

The next step is to sort the items in descending order by their annualized costs, with item Z at the top and item K at the bottom, as shown in Table 3-2. Next, calculate the cumulative cost, moving down the list from top to bottom. You will notice that the cumulative cost for each item is the sum of its cost and that of all the items above it.

The cumulative percentage of the total is then calculated for each item. For example, item Z represents 29 percent of the total (36,668.21 / 126,554.85 = 29 percent). Finally, you assign the ABC codes by applying the breakpoints depicted earlier. The *A*-items are those with cumulative costs less than or equal to 80 percent of the total annualized value; *B*-items are those with cumulative costs between 80 percent and 95 percent. The remaining parts are *C*-items.

SCM Team Exercise: Using the ZXY Company example, apply the Pareto 80/20 rule to each part in your own company's inventory to determine which ABC code to assign. Assign a code, with *A*-items representing the 80/20 stratum, *B*-items the middle 15/30, and *C*-items the bottom 5/50 stratum. (*Note*: Be sure to perform this exercise, as the SCM team will draw on its own ABC classification to complete many of the exercises in this workbook.)

Review the results of your ABC classification. Are any patterns or commonalties apparent? For example, are many of the *A*-items from the same product line or common to multiple products? Do you purchase them from the same supplier? Do they share a material commodity? Is unit cost or volume the determining factor, or is it a combination of the two? List any other patterns that jump out.

Notice in Table 3-2 that the classical breakpoints are not met exactly. For example, *A*-items actually comprise 19 percent of the parts and 78 percent of the value, not the precise breakpoint of 20 percent of the parts and 80 percent of the value. Close enough!

The team should also begin to think about how the costs and volumes on which ABCs are based might influence other inventory management decisions, such as lead-time and lot-size objectives and what levels of safety stock might be appropriate.

In the next chapter, we will build on these concepts and introduce an effective tool for depicting and measuring the material usage and replenishment cycle in a typical manufacturing environment.

CHAPTER FOUR

The Saw Tooth Diagram—
Analyzing Inventory Behavior

The SCM team is now ready to learn how to use a graphical analysis tool to depict the material usage and replenishment cycle that is typical in a manufacturing environment. This powerful visual tool, known as the *saw tooth diagram*, is used to analyze inventory behavior over time. In this chapter, you will learn about the formula and basic patterns of the saw tooth diagram. In Chapter 7, we will use the saw tooth diagram to analyze a number of real-world scenarios.

Saw Tooth Diagram Basics

The saw tooth diagram in Figure 4-1 depicts the inventory level for a typical part number, showing inventory declining over time as it is consumed (downward sloping lines) then jumping back up as replenishment orders are received (vertical lines).

The process begins with inventory at some level above order point (OP). As soon as inventory drops below the order point (called "breaking the order point"), a replenishment order is placed. Material continues to be consumed until the order is received. Some number of days (or other unit of time) after the order was placed, depicted by the lead time (LT), material is received, and inventory level returns to a point above order point.

The quantity of the replenishment order is the replenishment lot size (LS). Assuming that inventory is consumed at an even pace equal to the

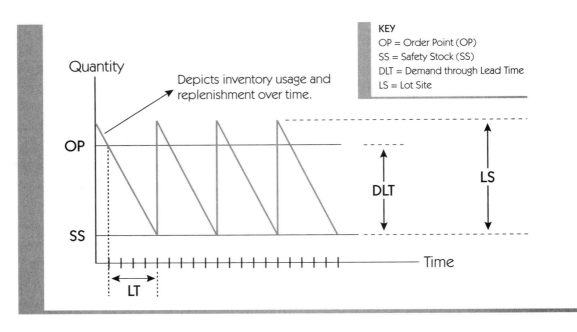

Figure 4-1. Saw Tooth Diagram Depicting the Inventory Level for a Typical Part Number

average usage rate, the inventory level will decline over time until it reaches the safety stock (SS) level. At that time the replenishment order is received, immediately raising inventory to the peak level. The amount of inventory used during the replenishment lead-time is called *demand-through-lead-time* (DLT).

A simplifying assumption of smooth, even usage makes the chart easier to understand. In reality, usage will sometimes be above average (a steeper decline) or below average (a shallower decline). You will need safety stock to protect against the high side variations. We will discuss techniques for determining how much safety stock is needed in a later chapter.

Lean Note: Excess inventory is one of the seven wastes in lean manufacturing. Inventory includes excess raw material, WIP, or finished goods, and exists at any stage of the process, causing additional wastes such as transport, storage, damage, and delay. It also ties up the company's resources: people, equipment, money, and energy. Although lean thinking considers inventory "waste," some amount of buffer inventory is necessary. The objective is to drive inventory to the lowest level required to maintain production flow and customer service.

Saw Tooth Example—Analyzing the Behavior of a Part Number

Figure 4-2 illustrates how you can use the formulas of the saw tooth diagram to analyze inventory behavior of a part number. The part number illustrated in the figure has annual usage of 1,500 pieces (5.95/day), a lead time from the supplier of 10 days, and a lot size of 100 pieces.

Sample Values

LT = 10 Days   Avg Days/Month = 21

Annual Demand = 1,500   LS = 100

SS = 25% of DLT

- **Average Daily Usage (ADU) =**
  Annual Sales / (12 months x 21 days/month) = 1,500 / 252 = 5.95
- **Demand through Lead Time (DLT) =** ADU x LT = 5.95 x 10 = 60 (Rounded)
- **Safety Stock (SS) =** 25% x DLT = 0.25 x 60 = 15
- **Order Point (OP) =** SS + DLT = 15 + 60 = 75

Figure 4-2. Example of Analyzing the Inventory Behavior of a Part Number

You calculate the demand-through-lead-time (DLT) as the number of pieces expected to be used, on average, while waiting for the replenishment order to be received. In this example, the DLT is 59.5, rounded to 60 pieces, and we assume management policy has set the safety stock level at 25 percent of DLT, or 15 pieces. Therefore, the order point can be calculated as SS + DLT = 75 pieces.

Figure 4-3 translates these calculations into the saw tooth diagram. Notice that the lot size of 100 pieces guarantees that when a replenishment order is received, the inventory will always return to a point above the order point. *This point is critical.* If the lot size is less than the order point amount, it is possible that inventory will remain below order point when the order is received. Even if another order is issued immediately upon receipt of the first order, there is an elevated risk of stock out. This situation is identical to placing an order late, that is, some time *after* breaking order point.

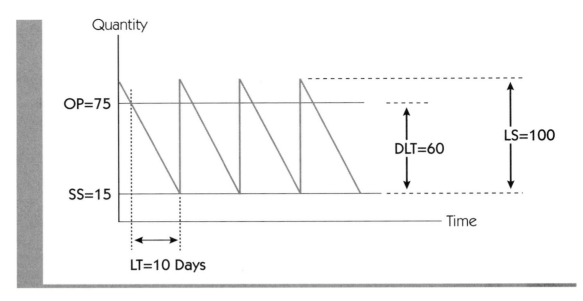

Figure 4-3. Saw Tooth Diagram from Figure 4-2 Calculations

> **SCM Team Note:** Later we will examine a replenishment model which uses multiple order triggers, with multiple open orders "in the pipeline" at any given time. The requirement for the lot size of an individual order to be greater than order point does not apply in that case. However, the combined lot sizes of all the individual orders must still meet this requirement.

Saw Tooth Assumptions vs. Reality

When drawing a saw tooth diagram, it is helpful to simplify the chart with the following assumptions:

- Demand is smooth, and occurs at a rate equal to average usage.
- Delivery of replenishment orders is predictable, occurring the lead-time number of days (or minutes, hours, etc.) after the order was placed in quantities equal to the lot size.
- Quality of parts on replenishment orders is 100%. All parts received are usable.

In reality, you can challenge all of these assumptions because from time to time they will not hold (see Figure 4-4).

The reason for this is that companies have little control or influence over their customers' buying patterns. However, as the customer to their sup-

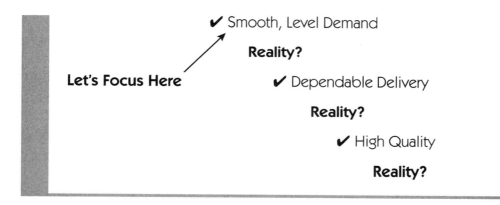

Figure 4-4. Challenge the Assumptions

pliers, companies have considerable control, up to and including terminating their relationship with a supplier or buying from another supplier if the first supplier does not meet delivery and quality expectations. In cases where your suppliers are not meeting these expectations, you can establish *temporary* safety stock levels while the supplier works to correct the underlying issues. However, this is not an acceptable long-term approach. In fact, as a rule, I recommend that <u>only demand variation be covered by safety stock</u>.

Lean Note: The long-term lean goal for inventory levels is to minimize buffer amounts of raw materials and WIP, to produce products in sync with customer orders, and to ship these orders as soon as the manufactured products are completed. Because of decreasing inventory on the floor, inventory turns are more rapid, dramatically reducing the need for floor space.

Analyzing Demand Pattern Variability

The saw tooth formula illustrated in Figure 4-5 shows what occurs when a part does not have a smooth and even demand. The part shown here was used only five times in the last year. Each time the part was used, a different quantity was needed. However, on average, 50 pieces were needed each time. Let's assume the part is inexpensive and management has decided to apply a 100 percent safety stock factor to protect against erratic demand.

Applying the standard calculations, DLT is 10 pieces. With a 100 percent safety stock factor, SS is also 10 pieces, giving an order point (OP) of 20 pieces. Is there a problem here? Yes! If inventory is at, or near, the OP of

Sample Values

LT = 10 Days   Part Used 5 times/year.

SS = 100% of DLT   Average of 50 pieces per time.

- **Average Daily Usage (ADU) =**
  Annual Sales / 252 (12 months x 21days/month) = 250 / 252 = 1 (rounded)
- **Demand through Lead Time (DLT)** = ADU x LT = 1 x 10 = 10 (Rounded)
- **Safety Stock (SS)** = 100% x DLT = 10
- **Order Point (OP)** = SS + DLT = 10 + 10 = 20 pieces.

Is There a Problem Here...?

Figure 4-5. Analyzing Uneven, Erratic Demand

20 pieces, when the next order arrives (average order quantity of 50), the probability is that there will not be enough stock to cover demand.

Potential Solutions for Covering Potential Stock Outs

There are a number of options for dealing with too little stock to cover demand (see Figure 4-6). Consider these alternatives:

- If the part is inexpensive and kanban is the desired material management approach, set OP at 50 (or higher) and establish a lot size sufficient to cover expected demand over the next 6 to 12 months. This ensures parts are available at a higher percentage of the time to cover expected demand.
- If the part is expensive, carrying just-in-case inventory may not be practical. If an MRP or similar system is available, consider putting this part under MRP control. This alternative requires that the part's demand is known or that you can accurately forecast it within the replenishment lead time.
- If the part is expensive and has a long lead time (longer than the customer's lead time for the finished product), it may be possible to negotiate a longer lead time for the finished product with the customer. This option would have to be coordinated closely with sales and marketing to ensure no negative customer issues result.
- Request a forecast from the customer, providing advance notice of future orders.

Potential Problem

- Potential Stock Out if order for 50 arrives when inventory level > 20, but < 50.

Potential Solutions

- Set Order Point at 50+ and Lot Size to cover 6–12 months. (Good if part is inexpensive.)

- Put part on order-as-needed (MRP-like) process. (Good if part is expensive and demand is known within replenishment LT).

- Set LT of product to cover combined replenishment and manufacturing time. (Potential issue with customers).

Figure 4-6. Potential Solutions for Stock-Out

Lean Note: In a nonlean environment, many companies plan for extra inventory (safety stock) to cover problems such as production imbalances, late deliveries from suppliers, defects, equipment downtime, and long setup times. In other words, inventory waste masks problems. The lean alternative is to "lower the water level" of inventory, exposing the "rocks" (problems), then systematically eliminate, or minimize, those problems.

SCM Team Exercise: Using the ABC classification you developed in Chapter 3, identify a part or item that is tied to a supplier and create a saw tooth diagram depicting its expected average demand and replenishment. Then compare the actual historical replenishment based on purchase order history. Discuss any differences between the two, possible reasons for the differences, and potential improvement opportunities.

Was actual demand for the part erratic? Were any supplier deliveries or quality issues evident during the selected time period? Were orders placed and received in sync with customer demand, or did inventory, once received, sit on the shelf for long periods before being consumed? Discuss how the historical pattern may have been different using kanban replenishment. Repeating this exercise for several parts, perhaps from different suppliers, will provide additional insights.

CHAPTER FIVE

Lead Time and Lot Size Guidelines

We now turn our attention to the supply chain or more specifically to parts and materials from suppliers that the kanban system should be managing. Although the focus will be on external suppliers, we will also discuss managing in-house manufactured parts. We begin by returning to the ABC classification discussed in Chapter 3 and developing some rules of thumb for lead times and lot sizes. Once we establish these rules, the SCM team can review some related points, such as packaging, frequency of delivery, quality expectations, delivery modes, and other factors that may affect lead time or lot size values.

For determining lot sizes, as a rule, follow the pattern based on the ABC criteria as illustrated in Figures 5-1 to 5-3 and discussed in Chapter 3. As you will recall, it is important to turn the *A*-items very quickly because they represent the highest inventory value. In other words, it is best to keep limited stock in house relative to usage and sales by ordering *A*-items very frequently and in small quantities. At the other end of the spectrum are *C*-items. Because these represent a small fraction of total inventory investment (approximately 5 percent), turning this inventory frequently may actually cost more in transportation and handling costs than the value of the inventory itself. Therefore, order *C*-items infrequently and in relatively large quantities. *B*-items would naturally fall somewhere in between the two extremes, so use some moderation in managing *B*-items.

You can further specify these ABC lot-size guidelines in terms of days-of-supply. For example, *A*-items might be replenished daily or weekly with

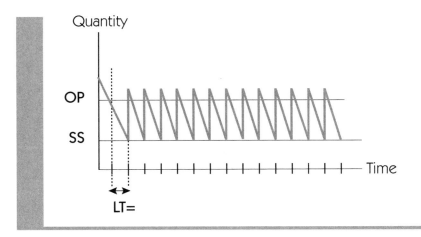

Figure 5-1. ABC Code *A*-Items (small, frequent orders)

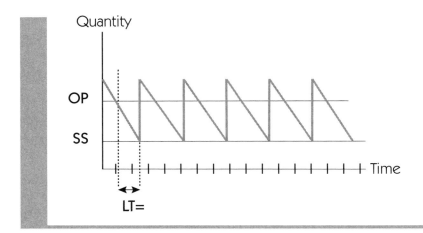

Figure 5-2. ABC Code *B*-Items (moderate, less frequent orders)

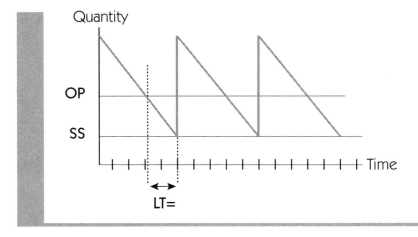

Figure 5-3. ABC Code *C*-Items (large, infrequent orders)

a lot size of 1–5 days of supply (i.e., a lot size sufficient to cover average demand over 1 to 5 days). *B*-items might be replenished monthly with a lot size of 20 days of supply. *C*-items might be replenished quarterly with a lot size of 60 days of supply. (In each case, the days represent manufacturing days not calendar days).

Such rules serve only as general guidelines. A better approach is to use a total cost model, such as the EOQ discussed below, and to compare the result of such a model to the general guidelines. Where dramatically different lot sizes result, review why the difference exists and decide on appropriate lot sizes taking into account any special circumstances presented by these parts. In other words, manage by exception. We will discuss the EOQ approach and examine some sample exception cases later in this chapter. First, let's take the next step in developing the Pareto Chart you created in Chapter 3.

> **SCM Team Exercise:** Return to the Pareto chart you created earlier to set ABC codes for part numbers in your company. Now add LT and LS values for each part number. Within each ABC category, discuss lot-sizing rules, whether formal or informal, that currently exist. Are any consistent patterns evident within ABC categories? Do you currently use any cost models to set lot sizes? What, if any, management by exception processes are in place?

Using the Economic Order Quantity (EOQ) Formula

A tried and true total cost model for determining cost-effective lot sizes is the Economic Order Quantity (EOQ) (see Figure 5-4). The EOQ formula takes into consideration the costs of placing an order (setup costs) as well as the costs of carrying the resulting inventory, referred to as *inventory carrying costs*. The *total cost* is the sum of these two individual costs.

A word of caution before we go further. The EOQ gives a lot size that minimizes total cost of a part. A key factor in total cost is the setup, or ordering, cost. In a lean environment, you continuously pursue setup cost reduction. As you reduce setups, you can also reduce lot sizes, driving continuous improvement in both operating costs and inventory turns. EOQ therefore provides a good short-term solution to minimize costs of the *current state*. But don't forget to revisit lot sizes regularly and capture potential savings (if any) if the cost elements change, especially as you reduce setup costs.

What is EOQ?

A formula to determine the order quantity that minimizes a part number's Total Cost, which is:

- Ordering (or setup) Cost
- Inventory Carrying Cost

Why use EOQ?

Because companies assume set-up costs are fixed, once calculated.

- OK to use in the short term to determine "current" minimum total costs.
- Challenge the assumption that fixed costs "are" the current set-up costs.
- Use to recalculate set-up cost periodically .

Figure 5-4. The "What" and "Why" of EOQ

The EOQ formula assumes a number of conditions, including stable, repetitive usage, and linear carrying costs, which don't necessarily hold up in reality. For example, Figure 5-5 illustrates the total cost along a continuum of lot sizes.

The first important point is that as lot size increases so does the carrying cost, because the company must carry inventory, whether purchased or produced, for a longer period of time before it is consumed. Secondly, the EOQ formula assumes carrying costs are incremental and linear, which ignores the reality that the cost curve is typically more of a step function because companies add or remove physical infrastructure in "chunks" rather than incrementally. However, the EOQ equation is quite forgiving and lot sizes within a broad plus/minus range of the EOQ produce favorable results, so applying the EOQ's assumptions of smooth, linear costs is not a major issue in practice. In fact, one of the main components of carrying cost is the cost of money. Because carrying cost ties up capital, companies correctly account for the cost of managing inventory like any other capital asset: as a linear percentage of total cost. Typically, they do this by considering it as the cost of borrowing money or account for it as a potential return lost by not investing in other areas. They then add a few percentage points to account for the internal structural costs.

Unit costs typically decline as lot size increases. In the case of manufactured parts, such as stamped metal or molded plastic parts, the cost of

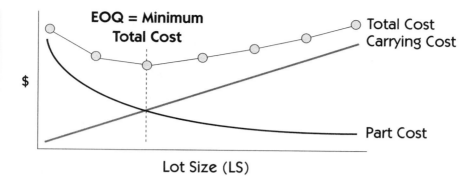

Part Cost

- Cost per part, including set-up and unit cost. Two types of parts:
  - Purchased Parts: Set-up = Ordering Cost
  - Make Parts: Set-up = Set-up Time X $/Hour
- **Cost Impact on Total Cost:** The greater the lot size, the smaller the part cost per unit.

Carrying Cost

- Cost of carrying parts in inventory.
- **Cost Impact on Total Cost:** The greater the lot size, the greater the inventory carrying cost.

Figure 5-5. Total Cost Along a Continuum of Lot Sizes

setup is amortized over all parts produced, creating the function shown as "part cost" on the graph. In the case of purchased parts, price breaks for higher quantities create a similar function, though not as smooth as the curve depicted by Figure 5-5.

> **Lean Note:** Excess inventory is one of the seven deadly wastes. It ties up the company's people, equipment, money, and energy. When inventory remains in the warehouse, there is no financial return on this investment. It also causes additional wastes in handling, transportation, storage, damage, and delay—again adding costs and diminishing the potential return on the company's capital.

The *total cost curve* is the sum of the two individual costs. This curve is at its lowest point where the two cost curves intersect, which is the

> **SCM Team Discussion:** The EOQ tends to fix a company's initial "current" setup costs, that is, once a company sets a lot size that minimizes the total cost, the tendency is to leave it alone. Challenge this short-term mindset by focusing on opportunities to reduce setup and ordering costs. Also, as you lower these costs, recalculate the EOQ to capture the new savings. What setup cost, or ordering cost, assumptions currently exist in your company? Are there opportunities to challenge these assumptions in the near term? What cost reduction potential exists?

minimum total cost or the EOQ. The EOQ formula in Figure 5-6 includes the following variables:

- annual usage expected for the part in question
- fixed cost of setting up for production or for placing an order
- inventory carrying cost (expressed as an annual percentage rate)
- unit cost of the part

For purchased parts, the unit cost is the cost of one unit (cost per piece, pound, square foot, etc.). For manufactured parts, the unit cost is the discreet cost of producing one unit of production, ignoring all setup costs. For example, if a part takes five minutes to produce and the hourly production rate is $50, the unit cost is $50/hour × 5 minutes / 60 minutes/ hour = $4.17. Figure 5-6 gives the cost of each variable, yielding an EOQ of 471 pieces.

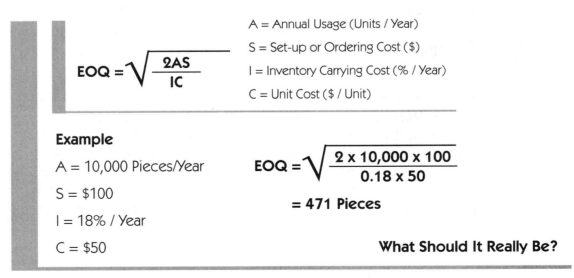

$$EOQ = \sqrt{\frac{2AS}{IC}}$$

A = Annual Usage (Units / Year)
S = Set-up or Ordering Cost ($)
I = Inventory Carrying Cost (% / Year)
C = Unit Cost ($ / Unit)

Example

A = 10,000 Pieces/Year
S = $100
I = 18% / Year
C = $50

$$EOQ = \sqrt{\frac{2 \times 10,000 \times 100}{0.18 \times 50}}$$

= **471 Pieces**

What Should It Really Be?

Figure 5-6. Adding A, S, I, C, Variables to EOQ

Now let's assume the part in Figure 5-6 is a supplier's part and the supplier is willing to give a price break of $10 for quantities above 1,000 pieces. Should you accept the lower price in exchange for buying in larger lot sizes? To answer, recalculate the EOQ with a unit cost of $40 ($10 off the original $50). The result is 527 pieces, well under the quantity required to obtain the price break. Therefore, accepting the price break in exchange for a lower unit price would actually increase your total cost for this part. (For a more complete understanding of this concept, refer to the EOQ graph and picture what the total cost might be far to the right of EOQ: e.g., 1,000 pieces vs. 527 pieces). In other words, to minimize total cost, the lot size should remain somewhere around 471 pieces, even though the unit cost (price) is higher than it would be with the higher-volume price break.

SCM Team Note: Always *consider* supplier quantity price breaks, but be aware that buying greater quantities at lower prices may not provide total cost savings. Be careful to analyze the total cost impact of quantity breaks before accepting them.

Fine-tuning the EOQ

As mentioned earlier, the EOQ formula is quite forgiving because the bottom portion of the total cost curve is relatively flat, so if you select a lot size reasonably close to the minimum (the exact EOQ), you will have a total cost that is still close to the minimum. It is therefore reasonable to round off the actual lot size to what is pertinent to company policy or other considerations that make sense, such as packaging multiples, days-of-supply guidelines, or raw material run-out quantities (see Figure 5-7). Using our earlier example (Figure 5-6), the EOQ of 471 pieces might be rounded to 500 if the supplier's standard packaging is in boxes of 250 each.

Another rounding approach might be contingent on achieving a desired number of days of supply. For example, you might want to purchase A-items on a weekly basis, allowing for a weekly milk run between the plant and a number of local suppliers. If average daily usage is 40 pieces (200/week), rounding to 250 (one standard package) might make sense. In the case of in-house manufactured parts, running out a raw material, such as a coil of steel or a full sheet of laminate, might eliminate scrap or the need to manage residual raw materials. The EOQ formula does not consider such cost factors, but a company should take them into account if the costs are significant.

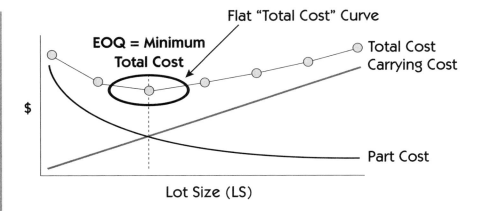

The minimum point on the Total Cost curve is relatively flat which:

- Provides Opportunity for SCM Team and Management Discretion in considering other factors with minimal cost impact.

- Can round the EOQ optimal point to something that "makes sense" for a particular company or situation, such as:

  - Container Quantity

  - Days-of-Supply Guidelines

  - Run-out a raw material

Figure 5-7. Fine Tuning the EOQ

In any of the above cases, if rounding changes the LS dramatically from the EOQ, you may need to take some other action. Do not get locked into thinking that current-state factors are fixed. A good lean principle is to replace "no because" thinking with "yes if" thinking. Examples might include establishing new packaging multiples or purchasing smaller coils or sheets of raw material. Replace "no, we can't use the new EOQ because packaging multiples are too large" with "yes, we can use the new EOQ if we establish smaller box quantities with our supplier."

The SCM team can also use the EOQ formula to test various alternatives. Continuing with our earlier example, if employing a shared milk run reduces transportation costs to $25, applying that reduction to the ordering cost and recalculating EOQ gives a lot size of 236 pieces, or roughly 6 days of supply (vs. 12 days of supply at 471 pieces). In this case, lowering the lot size to a weekly supply and employing the milk run is a good rounding alternative, since the final value of 5 days-of-supply is reasonably close to the EOQ (see Table 5-1).

Table 5-1. Employing a Milk Run

Fixed Values			Set-up Cost alternatives with resulting EOQ and #Days-of-supply values		
			S	EOQ	# Days
A	10,000	(Approx. 40/day)			
I	18%	Alternative 1	$100	471	12
C	$50	Alternative 2	$25	236	6

Setting Lot-size Guidelines

As discussed in the previous example, you might set lot-size goals in terms of number of days of supply. Figure 5-8 shows one possible set of parameters, based on ABC code. In general, ABC objectives recognize that *A*-items tie up the majority of inventory value and have a significant impact on the carrying costs of inventory.

> **SCM Note:** Though *B*-items have a lesser impact on inventory costs than *A*-items, they warrant some attention. Discuss their importance and seek out any possible cost opportunities. On the other hand, *C*-items have a <u>minimal</u> impact on cost. Order these parts infrequently, in fairly large quantities. The idea is to free up as much time as possible so the SCM team can concentrate on *A*-items.

In Figure 5-8, the desired lot size is 1,600 units. If the standard packaging is 500 per box, then 3.2 boxes are required to meet the lot size. The example lists three options or alternatives for implementing the desired lot size, including rounding the number of cards up or down and renegotiating packaging quantities with the supplier.

Observe one caution: When you use a single-card kanban approach, the *lot size must be at least as large as the order point.* Otherwise, if stock reaches zero before the order is received, the order point cannot be fully reset, creating the risk of another stock out (see Figure 5-8).

Continuing with our example (see Figure 5-9), let's assume you have renegotiated packaging quantities with the supplier, and the container size is now 400 pieces. To eliminate the stock-out concern, and to minimize

ABC	Frequency	Lot Size
A	Weekly	5 Days
B	Monthly	20 Days
C	Quarterly	90 Days

Example

- A-Item
- Average Daily Usage = 320
- Lead Time = 10 Days
- Container Qty = 500/Box

Lot Size Decision

- 5 Day Supply = 1,600
- No. Boxes = 3.2
- Options:
  - Set LS = 3 Boxes (1,500)
  - Let LS = 4 Boxes (2,000)
  - Request 4 Boxes of 400
- Caution: If LS (Days) < OP

  (Risk of stock out because replenishment order may not fully reset inventory above OP.)

Figure 5-8. One Possible Set of Parameters Based on ABC Code

Example

- A-Item
- Average Daily Usage = 320
- Lead Time = 10 Days
- Container Qty = 400/Box
- Vendor ships in pallet quantities, 4 boxes/pallet. (Vendor multiple equals 1,600 pieces.)

DLT + SS

- DLT = 320 x 10 = 3,200
- SS (10%) = 320
- OP = DLT + SS = 3,520
- If LS = 1,600 . . .
  - # Cards = OP / LS = 2.2 (Round up to 3)

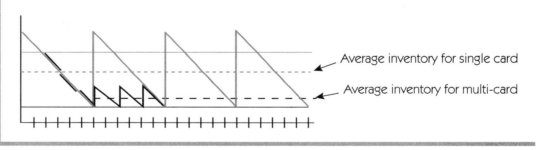

Average inventory for single card

Average inventory for multi-card

Figure 5-9. Lot Size Must be At Least As Big As the Order Point

inventory, you might apply a multi-card approach. To calculate the number of cards required, you first need to know the order point. Calculating the Order Point (DLT + SS), yields a value of 3,520 pieces. Assuming a lot size (and therefore a kanban card quantity) of 1,600 pieces, we see that three cards are required (3,520 / 1,600 = 2.2, rounded up to the next whole number of cards).

However, to maintain a single-card kanban system, the lot size would have to be at least 3,520 pieces (equal to the OP) to ensure the order point can be reset in the event of a stock out. Therefore, your lot size would also have to be at least 3,520 pieces. Since the vendor's lot-size multiple is 1,600, your lot size would have to be 3 pallets, or 4,800 pieces. The main weakness in this approach is with trying to manage multiple pallets with a single kanban card. As a general rule, every kanban container (in this case every pallet) has its own kanban card.

The multi-card approach yields lower average plant inventory as compared with a single, larger order, as depicted in the graph in Figure 5-9. For example, in this case using three cards, it is likely that only one card would be in the plant at any given time. The second would be associated with the order going to the supplier; the third would be in transit from the supplier to the plant. In this way, inventory will always be available when it is needed, without having to carry the full cost of inventory in-house. In addition, the multi-card approach assures that each pallet (or other packaging multiple) has its own kanban card. (We will revisit this point in Chapter 7).

> **Lean Note:** A lean inventory principle is to turn inventory rapidly through frequent deliveries of small lot sizes to approach the "goal" of one-piece flow. In the spirit of turning inventory, use the ABC classification to identify continuous improvement opportunities to develop small lot sizes.

Managing Replenishment Lead Times

Managing replenishment lead time is critical to an effective kanban system. Figure 5-10 outlines some basic lead-time principles. In most cases, do what is necessary to shorten lead times. For parts managed on kanban, long lead times require higher inventory levels, affect the ability to expedite delivery for unexpected demand and in general limit flexibility. Therefore, excessive lead times represent a form of *muda* (waste), which costs the company money.

For those suppliers whose internal operations do not facilitate short lead time response to orders, or those who have long setup, use the option of stocking agreements to ensure these suppliers always have enough material in stock to cover their own DLT and safety stock. Under such an arrangement, suppliers would stock a negotiated amount of inventory for your company, filling your company's kanban orders from stock and

Purchased Part Lead Times

- Smaller is better
- Lead time affects inventory
- Long lead time affects ability to expedite
- Early delivery is bad!

In-house Manufactured Part Lead Times (Make Parts)

- Lead time based on internal processes
- Ability to expedite usually greater than for purchased parts
- Set-up reduction drives improvements in lead time and lot size

Figure 5-10. Basic Lead Time Principles to Manage External and Internal Replenishment Lead Time

replenishing their own stock based upon internal processes. A supplier's lead time to you would typically become one day plus transportation, because the supplier would need only pick and ship to fill your kanban orders. (At a minimum, the supplier's stocking quantity is equal to your company's average daily demand times the supplier's total lead time to manufacture or acquire the component, plus an agreed-upon amount of safety stock). In return, the supplier may ask for contractual assurances to purchase this material to avoid being stuck with inventory that cannot be sold to other customers. Under such an arrangement, exercise care to ensure the company uses its contractually agreed upon material before engineering changes or mix changes significantly to reduce or eliminate the need for the parts.

> **SCM Communication Note:** In negotiating with suppliers, educate them on the principles of kanban, especially making it clear that kanban requires consistent lead times for you to achieve the goal of delivering to your customer. Also, encourage your suppliers to adopt kanban and pursue a lean/JIT program. For those suppliers that show interest, help provide kanban and lean/JIT training. In the end, the goal is to help suppliers use their demonstrated abilities to facilitate short lead-time response to orders.

Chapter 7 will cover the problem of suppliers who consistently deliver early and how this inflates inventory levels—not acceptable in a JIT environment. However, this situation presents an opportunity to review suppliers who consistently deliver early and work with them to help them reduce lead time on these parts. When reviewing delivery performance with your suppliers, use their history of early delivery as leverage to encourage them to reduce their lead times. Be cognizant of your suppliers' demonstrated abilities, reliability, and quality of work. Identify those suppliers you believe will achieve compliance with your company's needs. In the event a supplier is unwilling to reduce lead time, it is important to stress the need to deliver on-time (not early) and take steps to ensure compliance. If a supplier refuses to comply, you may have to use drastic measures, such as refusing early deliveries, imposing financial or performance measurement penalties, or (in the worst-case scenario), finding another supplier.

For in-house manufactured parts, lead times are typically based on internal manufacturing processes, including machine setups, run times, and interoperation move and queue times. It is sometimes easier to expedite in-house manufactured parts as compared to purchased parts; however, expediting impacts other manufactured parts in-process, or in queue, and adds cost. If you are expediting frequently, this also indicates internal waste and presents opportunities to use lean principles to find root causes and adopt countermeasures to improve your processes. The key to reducing both lead times and lot sizes for in-house manufactured parts is to use lean principles such as pit crews, poka-yoke, and other SMED (single minute exchange of die) principles to reduce set up times. Long setups drive large lot sizes (see the earlier discussion on EOQ), increase the time required to produce each lot, and thereby increase queue times between jobs, all of which tend to drive lead times higher. Higher lead times in turn increase plant inventory investment, tying up money that the company could use more profitably elsewhere.

> **Lean Note:** In lean thinking, lead time is the total time a customer must wait to receive a product after placing an order. One goal of lean is to continuously improve the lead times of the total supply chain, thus further improving the goal of one-piece flow. A long lead item at one of your suppliers (something they buy from one of their suppliers) could potentially shut you down if your supplier experiences a quality or delivery problem on that item. At a minimum, you must understand the total supply chain for the most critical parts in your operation.

Determining Lead-time Objectives

Every company needs to establish lot-size and lead-time objectives. This is especially true when implementing kanban. One option to assist the SCM team in determining lead-time objectives is to use the lead-time goals by ABC classification (see Figure 5-11). Generally, you want shorter lead times for *A*-items because these items account for the majority of a plant's inventory investment. Shorter lead times equal lower inventory. Because investment is lower for *B* and *C*-items, less aggressive goals may be set for these items. However, *B* or *C*-items deemed to be critical to the operation should have shorter lead times to ensure that you can expedite parts in the event of a stock out or other emergency.

Lead Time Goals by ABC

ABC	Lead Time Goal
A	< 5 days
B	< 10 days
C	< 20 days

Figure 5-11. Establishing Lead Time Objectives by ABC Classification

As with any generalized set of rules, there are exceptions. The SCM team will negotiate actual lead times with suppliers based on a number of factors, including location, price, and means of transportation, to name a few. Similarly, you work with manufacturing to ensure that lead times are in line with current manufacturing processes and delivery performance. As mentioned previously, setup reduction in manufacturing processes can significantly reduce lot sizes, queue times, and lead times, thereby driving lower inventory and greater flexibility throughout the manufacturing flow.

> **SCM Communication Note:** To ensure the company has consistent communication with all its suppliers, it is important that everyone responsible for implementing and maintaining the kanban system understands *the role of the supplier*. This workbook has cited several examples of when and where to communicate vital information to assure kanban's success. Everyone should know these touch points.

The purpose of these guidelines is to provide a quick, meaningful, way to identify parts whose lead times vary significantly from the objective, therefore highlighting opportunities for inventory reduction or improved flexibility. Again, use the management-by-exception approach with the guidelines providing the expected conditions against which to compare actual lead-time values.

SCM Team Exercise: Develop "Lead Time and Lot Size Goals by ABC" for your company. Then return to the ABC assignments completed earlier and apply the "should be" LT and LS values to each part. Compare actual LT and LS values to these objectives. If possible, use Excel or other software to check all part numbers and identify parts with significant differences between the current and "should be" values. (Any difference greater than 25 percent of the "should be" value might be a good starting point). Otherwise, select a few part numbers from each ABC category and manually review each. Discuss how to begin driving LT and LS values in the direction of the goals. What management or other department/team support do you need to drive these improvements?

CHAPTER SIX

Statistical Analysis 101 and Demand Variability

Companies can control many factors affecting inventory investment decisions, such as internal manufacturing processes and inventory replenishment techniques, but when it comes to the most important factor, the customer, they have little control. It takes a lot of work, as well as reliable data, to understand *your customers' demand patterns*. Understanding how *variability of customer demand* impacts your company and how to plan for it are key factors in providing reliable customer service. Nowhere is this more true than at the most basic level of planning and maintaining high levels of part availability. In this chapter, we will review a basic statistical model that supply chain professionals can use to analyze and quantify the amount of variability in a parts-usage pattern over time. Determining the amount of variability in a parts-usage pattern is the single most important thing supply chain professionals can do to establish the amount of safety stock required on a part.

Customer Demand Variability

In Chapter 4, I recommended that you establish safety stock levels to cover demand variability. But how does customer demand variability correlate to safety stock? The simple answer is that the more variable the demand pattern on a part, the more safety stock is required. Creating a model allows you to standardize your safety stock decisions tied to demand variability. More importantly, you can continuously use the model to track changing behavior (variability). There are several alternatives and tools for achieving this. One alternative is to set safety stocks based on high-level management decisions, such as the ABC rules presented earlier. In general, *A*-items tend to be more stable and require less safety stock than more-variable *B*- and *C*-items. However, a much more powerful and accurate technique is to apply a statistical model to each part's unique demand pattern. Figure 6-1 defines several statistical terms we will be using in this chapter.

- **Mean:** The arithmetic average of a group of values.

- **Deviation:** The difference, usually in absolute difference, between a number and the mean of a set of numbers.

- **Standard Deviation:** A measure of dispersion of data or of a variable. The standard deviation is computed by finding the differences between the average and actual observations, squaring each difference, summing the squared differences, dividing by n-1 (for a sample), and taking the square root of the result.

- **Normal Distribution:** A particular statistical distribution where most of the observations fall fairly close to one mean, and a deviation from the mean is as likely to be plus as it is to be minus. When graphed, the normal distribution takes the form of a bell-shaped curve.

Figure 6-1. Basic Definitions. Source: APICS Dictionary, 9th Edition, 1998

> **SCM Team Discussion:** Go over Figure 6-1, discuss the definitions, and determine the depth of statistical knowledge of the team. For example, has any individual used software such as Excel or other programming tools to calculate or model these statistics? Use this opportunity to make sure everyone adequately understands these terms and to bring those individuals who are less experienced in this area up to speed.

One basic assumption a company can make about demand variability is that demand is normally distributed around an average, or *mean*, value. Though not always a precise method, modeling demand behavior based on a *normal distribution* is much easier to do than using complex statistical techniques. This simpler approach has the benefit of being readily grasped and accepted by analysts, managers, and others. Figure 6-2 shows a normal distribution also known as a *bell curve*.

In a bell curve, the highest number of observations occurs at, or near, the <u>mean</u>. In Figure 6-2, the mean is at 13. Notice that the shape of each half

Statistics 101

- Mean is the average of all observations.

- Standard (Std.) Deviation is a measure of variability.

- +/- 1SD = 68.3%

- +/- 2SD = 95.5%

- +/- 3SD = 99.7%

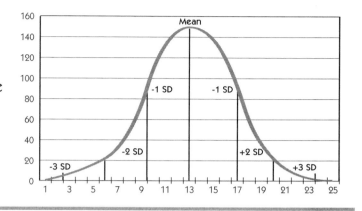

Figure 6-2. Example of Normal Distribution

of the curve is the mirror image of the other, indicating that there is an equal number of observations on either side of the mean, and the values are equally spaced away from the mean. The *standard deviation* is the key measure of how widely distributed these observations are. In a normal distribution, 68.3 percent of all *observations* occur within +/–1 standard deviation from the mean, 95.5 percent within +/–2 standard deviations, and 99.7 percent within +/–3 standard deviations. So how can you use the normal distribution to determine safety stock levels? The answer lies in a statistical device know as the Z score.

Using Z Scores

The *Z score* is a commonly used statistic associated with a normal distribution. This is similar to the percentages discussed above and illustrated in Figure 6-2. However, the Z score measures the percentage of observations from the left-most end of the distribution, up to some number of standard deviations from the mean. A negative Z score indicates a position to the left of mean; a positive Z score indicates a position to the right of mean. For example, in Figure 6-3 a Z score of 2 indicates two standard deviations above (to the right of) the mean. The Z table shows the correlation between various Z scores and the equivalent percentage value. In this case, a Z score of 2.05 is equivalent to a 98 percent probability that a particular value will be included in the shaded portion of the curve.

For purposes of setting safety stock, <u>Z score is the appropriate statistic to use</u>, and *should be applied to the variability of daily DLT values* not to

Z score measures the percentage of observations from the left-most end of the distribution up to some number of standard deviations from the mean.

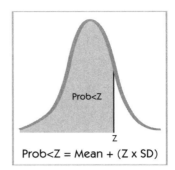

Z Prob	< Z
-0.67	25%
-0.25	40%
0.00	50%
0.67	75%
1.28	90%
1.64	95%
1.88	97%
2.05	98%
2.33	99%
2.58	99.5%
3.09	99.9%

Prob<Z = Mean + (Z x SD)

Example:

- Mean DLT = 10, SD = 2

- Service Goal = 98%

- Choose Z = 2.05

Set OP = 10 + (2.05 x 2) = 14.1 (Round up or down as desired)

Figure 6-3. Correlation Between Various Z Scores and the Equivalent Percentage Value

discreet daily usage values. Recall that demand through lead time (DLT) is the amount of inventory expected to be used during the replenishment lead time. Figure 6-4 shows an example of a daily DLT calculation based on 30 days and shows the number of pieces used for a particular part number. (30 days is used for illustration purposes only. In general, it is best to use at least one year of data to adequately model the variability of a part's usage.) Assuming a 5-day replenishment lead time, discreet daily DLT values are calculated. For day 1, the DLT is the sum of days 1 through 5. For day 2, days 2 to 6 are summed; for day 3, days 3 to 7 are summed; and so on. Then calculate the mean and standard deviation of the *DLT values*. The order point is based upon these statistical values. Figure 6-4 shows the results, with daily usage depicted by diamonds and DLT values by squares. Average DLT is the horizontal line at about 2,600.

Observe the similarity between the formulas for Z scores and our earlier order point formula, both shown below. Note that usage cannot be less than zero, so the left-most value on the curve is zero.

Z Score: Prob<Z = Mean + (Z x SD)

"Prob<Z" represents the probability of an observation being within the portion of the curve between zero and Mean + Z Std. Deviations.

Order Point: (OP) = Mean DLT + (X percent x DLT)

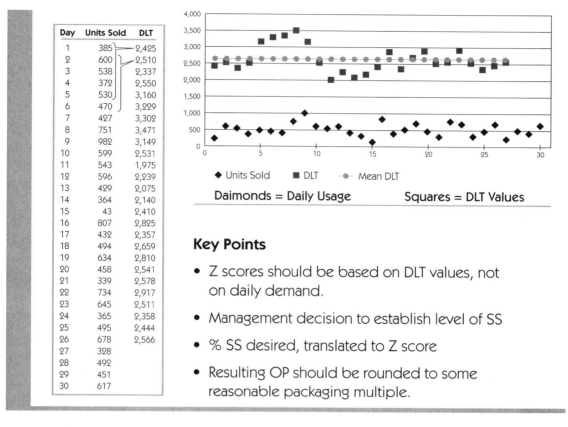

Day	Units Sold	DLT
1	385	2,425
2	600	2,510
3	538	2,337
4	372	2,550
5	530	3,160
6	470	3,229
7	427	3,302
8	751	3,471
9	982	3,149
10	599	2,531
11	543	1,975
12	596	2,239
13	429	2,075
14	364	2,140
15	43	2,410
16	807	2,825
17	432	2,357
18	494	2,659
19	634	2,810
20	458	2,541
21	339	2,578
22	734	2,917
23	645	2,511
24	365	2,358
25	495	2,444
26	678	2,566
27	328	
28	492	
29	451	
30	617	

Daimonds = Daily Usage   Squares = DLT Values

Key Points

- Z scores should be based on DLT values, not on daily demand.

- Management decision to establish level of SS

- % SS desired, translated to Z score

- Resulting OP should be rounded to some reasonable packaging multiple.

Figure 6-4. Daily DLT Calculation Based on 30 Days Showing the Number of Pieces Used for a Particular Part Number

In both cases, some amount of inventory is added to average (mean) DLT; this additional amount is your safety stock. The benefit of the Z-score formula is that it considers the specific statistical variability of each part number, rather than assigning some generalized percentage safety stock to a family of parts.

The Z-score formula represents the probability that an observation (amount of material needed at a point in time) is less than the amount of available inventory (mean plus some percentage based on Z score) at that time. In other words, it represents the probability that enough inventory will be on hand to cover expected demand. In the example in Figure 6-3, an order point of 14 assures that 98 percent of the time, the order point will cover the total demand for that part during the replenishment lead time.

In Figure 6-5, this 30-day DLT example shows the calculated mean and standard deviation of DLT values, as well as the results used in the Z-score formula to determine the order point. (Note: This example does not present detailed calculations of mean and standard deviation. To do this, you

> **SCM Team Note:** One key point bears repeating. When using the Z-score approach, base the statistics (mean and standard deviation) upon DLT values, not on daily demand. A simplified technique, producing the same result, will be presented later. For now, it is enough to understand the concept.

can use an off-the-shelf spreadsheet software, such as Excel, or other programming tools. Figure 6-1 describes the underlying calculation process). The average demand through lead-time is 2,618 units. Applying the Z-score formula yields a safety stock of 1,001 units, for a total order point of 3,619. Therefore, an order point of 3,619 will ensure that 99.5 percent of the time no stock outs will occur during the replenishment lead-time. Finally, it is reasonable to round the resulting order point to some convenient packaging multiple in order to simplify physical stocking and material handling.

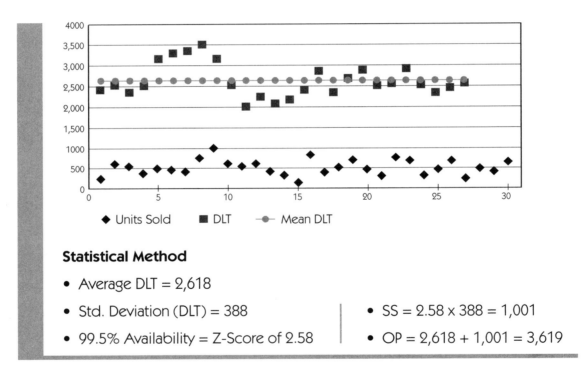

Statistical Method

- Average DLT = 2,618

- Std. Deviation (DLT) = 388

- 99.5% Availability = Z-Score of 2.58

- SS = 2.58 × 388 = 1,001

- OP = 2,618 + 1,001 = 3,619

Figure 6-5. Calculated Mean, Std. Deviation of DLT Values, and Results Used in the Z-Score Formula to Determine OP

Recommended Formula

In the previous SCM Team note, I mentioned a simpler calculation to determine safety stock. Rather than calculating daily DLT values, you can apply the following formula to the initial daily usage values to arrive at the same solution:

SS = Standard Deviation of Daily Usage x Square Root of LT x Z score

In our example, the result would be:

SS = 176.25 x SQRT(5) x 2.58 = 1,017

OP = 2618 + 1017 = 3,635

Note that the result is very close to the original daily DLT Z-score result. The reason the two results are not identical is that the variability is not exactly "normal"—that is, it does not exactly fit a bell curve. However, since you will most likely be rounding the calculated result to some convenient multiple, the small difference is meaningless in actual application.

Table 6-1 provides a convenient correlation of Z scores to service level for you to use in determining actual safety stock levels for your company.

Table 6-1. Z-Score Table (Correlation between Z-score and product availability [service level])

Z	%	Z	%	Z	%	Z	%
0.00	50.00%	1.00	84.13%	2.00	97.72%	3.00	99.865%
0.05	51.99%	1.05	85.31%	2.05	97.98%	3.05	99.886%
0.10	53.98%	1.10	86.43%	2.10	98.21%	3.10	99.903%
0.15	55.96%	1.15	87.49%	2.15	98.42%	3.15	99.918%
0.20	57.93%	1.20	88.49%	2.20	98.61%	3.20	99.931%
0.25	59.87%	1.25	89.44%	2.25	98.78%	3.25	99.942%
0.30	61.79%	1.30	90.32%	2.30	98.93%	3.30	99.952%
0.35	63.68%	1.35	91.15%	2.35	99.06%	3.35	99.960%
0.40	65.54%	1.40	91.92%	2.40	99.18%	3.40	99.966%
0.45	67.36%	1.45	92.65%	2.45	99.29%	3.45	99.972%
0.50	69.15%	1.50	93.32%	2.50	99.38%	3.50	99.977%
0.55	70.88%	1.55	93.94%	2.55	99.46%	3.55	99.981%
0.60	72.57%	1.60	94.52%	2.60	99.53%	3.60	99.984%
0.65	74.22%	1.65	95.05%	2.65	99.60%	3.65	99.987%
0.70	75.80%	1.70	95.54%	2.70	99.65%	3.70	99.989%
0.75	77.34%	1.75	95.99%	2.75	99.70%	3.75	99.991%
0.80	78.81%	1.80	96.41%	2.80	99.74%	3.80	99.993%
1.85	80.23%	1.85	96.78%	2.85	99.78%	3.85	99.994%
0.90	81.59%	1.90	97.13%	2.90	99.81%	3.90	99.995%
0.95	82.89%	1.95	97.44%	2.95	99.84%	3.95	99.996%

> **SCM Team Exercise:** Take a few moments to discuss the Z-score formula. Go over the example in Figures 6-4 and 6-5. Have each team member change a few of the variables in the figures, recalculate the Z scores, and then discuss the outcome and any difficulties. Apply both techniques, using the daily DLT values, and the simplified daily usage approach. Have the team again review the formulas and results. How close were the two techniques? If significantly different, what may have accounted for the differences? The purpose of this exercise is to clearly understand and know how to use the Z-score formula.

How Much Safety Stock Is Enough?

Now that we have fully developed the statistical model, we must ask, "How much safety stock is enough?" This is largely a management decision, and the guidelines offered in Figure 6-6 and the discussion that follows represents conventional wisdom on that question.

So How Much Safety Stock Is Enough?

- SS by ABC. Demand patterns are not the same for all parts in an ABC category.
- Statistical Analysis accounts for those differences.
  - Use Z Score

Combining these two ideas:

ABC	Z Score	SS Level
A	1.5	Low
B	2.0	Medium
C	2.5	High

Figure 6-6. Guidelines for Determining Safety Stock

Recall from the earlier discussion of ABC classification, that the intent of assigning ABC codes is to ensure that there is ample inventory on C-items to allow the SCM team to have a greater focus on A- and B-items. In other words, assign a higher proportion of safety stock to C-items. Because C-items represent only 5 percent of total inventory investment, carrying extra safety stock on these items requires minimal investment and frees up inventory control personnel to better manage the 20 percent of A-items, affording the greatest opportunity for inventory savings.

When applying these concepts, it might be reasonable to provide nearly 100 percent coverage of the C-items. In the example shown, a Z score of 2.5 provides 99.7 percent coverage. However, because of the higher value (cost) on A-items, the company may not be able to afford the same level of coverage. Choosing a 90–95 percent range of coverage may be sufficient for these items, represented by the Z score of 1.5.

It is reasonable to expect, then, that the SCM team will have to give special attention to A-items, and, to a lesser extent, B-items, to ensure that on those occasions when the order point does not cover all demand, you have special steps in place to obtain the necessary materials. For example, maintaining expediting, air shipping, and quick-ship agreements with suppliers can ensure that you avoid shortages in these cases. The SCM team should be *prepared and expect to take* such steps; the compromise for carrying an overall lower level of inventory. Properly managed, the

---

SCM Team Exercise: Using your ABC classification, choose a few parts and practice applying the Z-score formula to the daily demand history for each part. Choose at least one item from each ABC class. (Note: You will need to obtain daily usage data for each selected part which is often available from your computer system's transaction history, but you may need the IT organization's help in obtaining it). Remember for purposes of setting safety stock, Z score is the appropriate statistic to use, and, *in theory, should be applied to the variability of daily DLT values* not to discreet daily usage values. However, the simplified approach using the square root of lead time and daily-usage values provides the same result and is much simpler to calculate and to automate. Using Excel, or other available software, graph the daily usage and daily DLT values as shown in Figure 6-5. Does the data appear to be normally distributed? Or is it "skewed" to one side of the mean—i.e., are there more data points on one side than the other, and does the spread appear to be greater on one side? Draw a line on your graph at the recommended order point level (SS + Avg. DLT). How many daily DLT values exceed the line? Each point above the line represents a historical event where a stock out may have occurred. Discuss possible alternatives to deal with these events.

reward is a lower total cost of providing high-customer service rather than relying solely on the "just-in-case" mentality of maintaining high levels of safety stock.

Deciding Which Parts to Put on Kanban

One other key decision remains—whether or not to place a part under kanban control. As I mentioned previously, kanban works best when applied to stable, predictable demand. However, lower-cost or lower-volume products which have relatively erratic demand may still be candidates since they can be protected with safety stock at very little cost, and this cost will likely be less than ordering more frequently on an as-needed basis.

Using the simple statistical values of mean and standard deviation, we can analyze the relative variability between different part numbers. Dividing the standard deviation by the mean yields a value known as the Coefficient of Variability (CoV). A small CoV value, say in the range of 0 to 0.1, indicates very little variability; larger values indicate more variability.

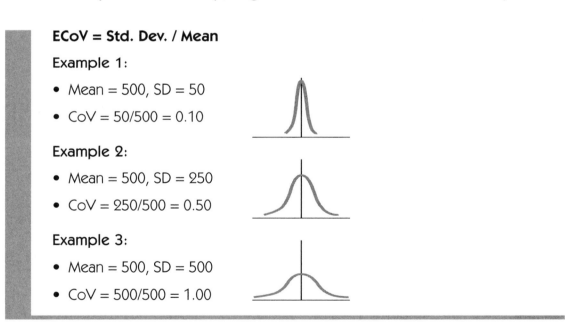

ECoV = Std. Dev. / Mean

Example 1:

- Mean = 500, SD = 50
- CoV = 50/500 = 0.10

Example 2:

- Mean = 500, SD = 250
- CoV = 250/500 = 0.50

Example 3:

- Mean = 500, SD = 500
- CoV = 500/500 = 1.00

Figure 6-7. Coefficient of Variability

Figure 6-7 shows three parts, each with the same average usage: 500 pieces. However, the variability differs significantly between part numbers, as indicated by the different standard deviation values. The differences are shown graphically to illustrate the point further. Example 1 shows a part with very little variability, while Example 3 shows a part with high variability. Next, we'll consider some possible decisions regarding how to manage these parts.

Example 1:

- Mean = 500, SD = 50
- Cv = 50/500 = 0.010

Low Variability Decisions:

- Minimal SS required
- Good candidate for kanban
- Consider automatic ordering

Example 2:

- Mean = 500, SD = 250
- Cv = 250/500 = 0.50

Moderate Variability Decisions:

- Moderate SS required
- Good candidate for kanban

Example 3:

- Mean = 500, SD = 500
- Cv = 500/500 = 1.00

High Variability Decisions:

- High level of SS required
- If low-cost, kanban still OK
- If high-cost, pursue means of reducing variability

Figure 6-8. Replenishment Alternatives Based on Variability

Figure 6-8 looks at each example again, this time examining some possible replenishment alternatives. Example 1 indicates a part with minimal variability. Minimal safety stock is required. Kanban would work quite well for this part; however, if the part is used very frequently, every day for example, it may be better to employ some type of automatic daily ordering technique to protect against disruptions in production. Say, for example, 500 pieces of this part are used daily, with an order point of 550 (500 pieces plus 50 for safety stock). One order should be placed each day, with multiple open orders in the pipeline at any given time to cover the supplier's lead-time.

Now, what if a problem occurs on the line, shutting it down for one day? If no products are produced, no parts will be pulled, and no kanban orders will be placed. A disruption in supply will occur lead-time-days later as a result of the delayed order. This will again disrupt production, and the situation is likely to recur.

An alternative is to schedule daily replenishment of this part, regardless of whether the parts were used or not. Under this approach, material is guaranteed to be available, however, care must me taken to temporarily turn off additional orders if inventory exceeds some predetermined maximum level. The Min/Max scheduling technique described in Chapter 11 can be used in these cases.

Example 2 indicates a part with moderate variability. Parts in this category are excellent candidates for kanban. A moderate level of safety stock, combined with kanban's manual-visual controls, will prove very effective in ensuring material availability at a reasonable inventory cost.

Example 3 indicates a part with high variability. A high level of safety stock would be required by kanban, making this approach expensive, especially for high-cost parts. In such cases, MRP or some other order-as-needed approach may be preferable. Inexpensive C-items in this range are probably still good candidates for kanban. The final decision is largely a judgement call, based upon whether or not other ordering mechanisms are available and the cost or complexity of using them.

The SCM team should now have a good grasp of the statistical analyses needed to determine which parts should be put under kanban control and how to calculate the safety stock and order points for those parts. In Chapter 11, we will return to these statistics to help determine appropriate Min/Max control limits for level-loading production.

CHAPTER SEVEN

Saw Tooth Exercises

Now that the basics of kanban, ABC classification, and the statistical safety stock model are in place, and the SCM team has begun applying these techniques to their own data, we can examine some common demand patterns and review ways to manage inventory by utilizing kanban. We will use the following patterns for this discussion:

- Decreasing demand
- Use of multiple kanban cards for frequently used parts
- Erratic demand

In each case, we will use the saw tooth diagram to illustrate the demand patterns and resulting impact on inventory.

Managing Kanban During Decreasing Demand

There are a number of factors that can drive decreasing demand, including the following:

- Economic downturns
- Competition eroding market share
- Life cycle patterns

Regardless of the cause of decreasing demand, it is important that the SCM team actively manage kanban to ensure they maintain inventory at the proper level for each part's actual or anticipated demand pattern. Figure 7-1 depicts the current state for a sample part using a single kanban card. In this model, when an order is received, inventory rises to a level above the order point, and a new order is placed only when the order point has been broken again (i.e., when inventory falls below the order point).

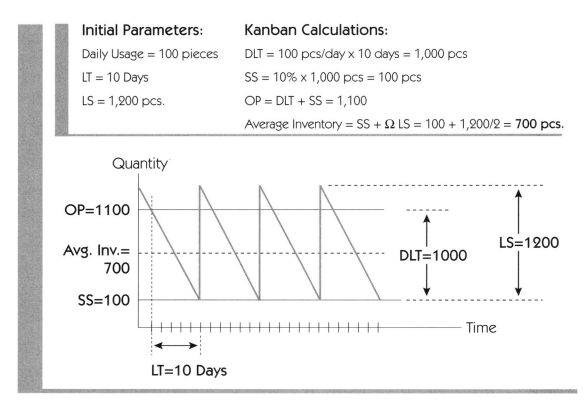

Initial Parameters:

Daily Usage = 100 pieces

LT = 10 Days

LS = 1,200 pcs.

Kanban Calculations:

DLT = 100 pcs/day x 10 days = 1,000 pcs

SS = 10% x 1,000 pcs = 100 pcs

OP = DLT + SS = 1,100

Average Inventory = SS + Ω LS = 100 + 1,200/2 = **700 pcs.**

Figure 7-1. Current State for Part Using Single Kanban Card

Applying the standard demand through lead-time (DLT) calculations discussed earlier and using a simple percentage safety-stock approach, yields an order point of 1,100 pieces with SS of 100 pieces. Now let's assume demand decreases to half the original level, now at 50 pieces per day. If the kanban level (order point) is allowed to remain at 1,100 pieces, applying the kanban order point, safety stock, and average inventory formulas leads to a new average inventory of 1,200 pieces, up from the original 700 pieces. (see Figure 7-2).

How did this happen? The answer is simple. When the order point is broken, an order is placed; however, demand is now half what it was, so that 10 days later you have only used 500 pieces rather than the 1,000 pieces originally expected. Therefore, when the order arrives, the inventory level

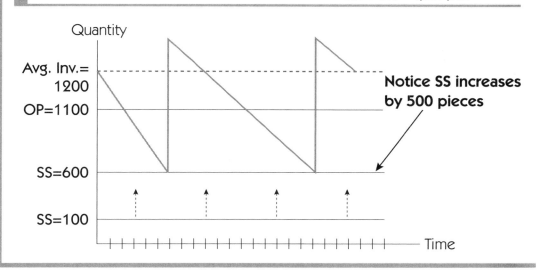

Assume 50% Decrease in Demand to 50 pieces/day

If no action is taken (order point allowed to remain at 1,100), the following will occur:

1) DLT = 50 pcs/day x 10 days = 500 pieces   2) Formula: OP = DLT + SS

3) SS = OP – DLT = 1100 – 500 = 600 pcs   4) **Avg. Inventory** = SS + ½ LS = 600 + 600 = 1,200 pcs.

Figure 7-2. Calculating 50 Percent Decrease in Demand for Part Using Single Kanban Card

is higher than originally planned. In effect, your actual safety stock rose by 500 pieces. Figure 7-2 shows the results of applying the decreased demand to the current state calculations: Inventory, on average, "bottoms-out" at a much higher level than originally planned. The increase is, of course, the result of the reduced DLT, driven by reduced demand.

You can see this effect mathematically by solving the order point formula (OP = SS + DLT) for safety stock. Since OP is already established at 1,100 pieces, and DLT is known (50 pieces/day = 500 pieces during the 10-day replenishment lead time), we see that safety stock is now 600 pieces:

OP = DLT + SS

Therefore: SS = OP – DLT

SS = 1,100 – 500 = 600 pieces.

Recommend Actions for Adjusting to Decreasing Demand

To prevent, or at least minimize the impact of demand changes on kanban levels, the team must take steps to plan for such changes and adjust kanban levels in advance of the change if possible. This means using a reliable forecast to plan demand changes and adjust kanban levels prior to the change. If such a forecast is available, it should also include overall trends (are sales expected to grow or shrink, and by how much?), and seasonality (which months or quarters tend to be high or low, and by how much?) Apply this information only to the DLT portion of the order point calculation. The safety stock should still be set based upon historical variability, as discussed in Chapter 6. Figure 7-3 shows some recommended actions for countering or preempting decreasing demand.

Recommended Actions

- Anticipate demand changes if possible.

- Analyze kanban levels regularly.

- Have mechanism in place to trigger OP change if impact > $xxx or < nn%.

- Focus on A-Items first. These have the greatest impact.

- If change in demand affects mix significantly, consider updating ABCs.

- Renegotiate Lot Sizes (major impact on average inventory).

Figure 7-3. Recommended Actions for Decreasing Demand

Whether or not the SCM Team can anticipate demand changes, they must have a process in place to identify changes quickly as they occur, as well as mechanisms to trigger order point changes when current kanban levels differ from required levels. You can measure the difference in a variety of ways, such as by percent, dollars, units, etc. The key is to *manage by exception*, with mechanisms in place that quickly identify any out-of-balance conditions.

For example, if the current kanban level for a particular part is less than the newly calculated DLT + SS, and the difference is greater than 10 percent (or whatever exception limit is deemed appropriate), increase the part's kanban level. A control limit of this type ensures that you will adjust up kanban order points that are *too low* before the safety stock level is eroded. Conversely, if a part's order point is high compared to the newly calculated DLT + SS, and the difference is greater than some agreed-upon limit, then adjust the order point down. In this case, a dollar-limit may be more appropriate, because the value of the inventory reduction to be achieved with a lower order point must cover at least the cost of making the change (labor, printing new cards, updating system records, etc.). Different exception limits may be set based on ABC because the percentage of safety stock, and the cost of material in each ABC strata, can be significantly different.

Again, the team should act on *A*-items first because they have not only the greatest financial impact but are typically the fastest-moving items and are thus more likely to create problems in manufacturing. Other actions may include recalculating ABCs (especially if the demand change effects product mix) and revising lot sizes to keep them in synch with inventory targets. We'll examine this last point in more detail next because it has the most direct impact on the supply chain.

> **SCM Communication Note:** Always keep kanban levels in-sync with demand. The team must regularly recalculate required OP levels, and make adjustments when current kanban order points are either too large or too small. In cases where OP changes effect lot sizes, you must negotiate with suppliers any changes in lot size, including any necessary modifications to packaging. If possible, partner with your suppliers regarding variability issues. For example, share your short term forecasting. Also develop possible contingency plans for sudden changes in demand. Communicating and developing appropriate responses to erratic demand with your suppliers will give you (and them) the agility to manage any demand changes quickly.

Applying the Solution for Decreasing Demand

Applying the management-by-exception process to our earlier example (Figure 7-2), it is clear the current OP is too large. Figure 7-4 provides a two-part solution, step one of which is to adjust the order point down to 550 pieces. The net effect is a reduction in safety stock of 50 pieces and a new average inventory level of 650. Note, however, that the peak inventory level is much higher than the order point, creating a long delay between orders and a high average inventory level. By reducing the lot size (order quantity) you can achieve a significant reduction in inventory while maintaining a high level of material availability.

Step 2 in Figure 7-4 shows the effect of reducing lot size to 600 pieces. Average inventory has now dropped significantly, from 650 pieces to 350. Figure 7-5 shows the final results of our efforts to reset kanban to a level consistent with the reduced level of demand (i.e., the new DLT + SS). Notice that there is a single reorder point and only one open order at

Applying the Solution

Step 1: Revise the Order Point

- DLT = 50/day x 10 days = 500 pcs.
- SS = 10% x 500 = 50 pcs.
- OP = SS + DLT = 550

Step 2: Reduce LS 50%

- New LS = 600
- New Avg. Inv. = SS + 1/2 LS = 50 + 300 = 350

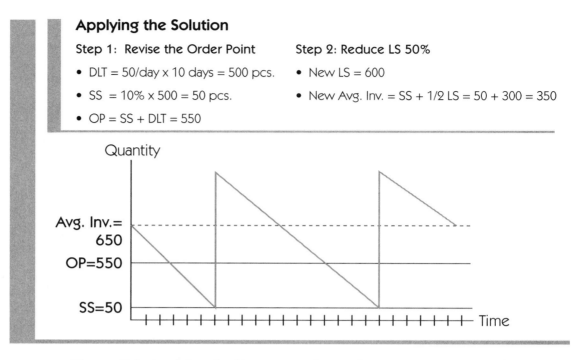

Figure 7-4. Applying the Two-Part Solution for Decreasing Demand

Reduction in Average Inventory from Smaller Lot Size

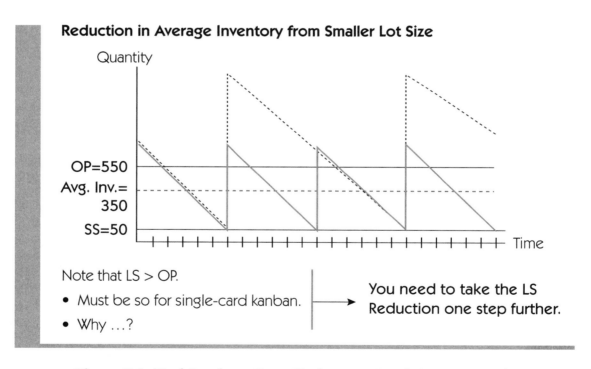

Note that LS > OP.

- Must be so for single-card kanban.
- Why …?

You need to take the LS Reduction one step further.

Figure 7-5. Final Results to Reset Kanban to a Level Consistent with
the Reduced Level of Demand

any given time. Recall from Chapter 4 that with this single-kanban approach it is vital that the lot size be large enough to raise inventory above the order point when an order is received. In other words, the LS must be greater than OP. (This is why a lot size of 600 was selected). If not, when the part is received, inventory may still be below OP and a future stock out is likely. As you will see in the next section, however, you can take the LS reduction one step further.

Further Lot-size Reduction is Possible Using Multi-card Approach

Further LS reduction is possible, with corresponding reductions in average inventory. This approach involves having multiple, smaller orders in the pipeline. Figure 7-6 depicts a multi-card approach in which multiple small orders replace the single large order discussed thus far. This approach requires that the supplier deliver more frequently and in smaller quantities.

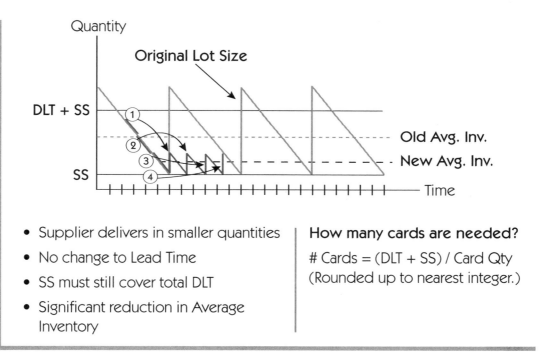

Figure 7-6. Multi-Card Approach That Uses Multiple Small Orders to Replace the Single Large Order

For example, you could divide an order of 1,000 into four kanban cards, each triggering an order of 250 pieces. The kanban system would require four containers, each with a kanban card attached. As each container is opened, its kanban card is pulled, and an order is placed. Lead time number-of-days later that order is received, yielding the replenishment pattern depicted by the small saw tooth patterns in Figure 7-6. Notice that the first order of 250 is triggered near the original OP. Let's say the lead time is 5 days. Five days after placing the order, 250 pieces are received, creating the first small saw-tooth spike. Shortly after this first order was placed, more parts were needed, and the second box was opened, triggering the second 250-piece order. This repeats as each subsequent box is opened. Five days after placing each order, the parts are received, creating the small saw tooth pattern. Notice the dramatic reduction in average inventory with this approach.

Two points must be noted. First, the calculated order point (DLT + SS) must always cover the full demand through lead time plus whatever safety stock is deemed necessary. Second, lead time has not been altered, although it may appear so by looking at the time between receipts. (In the previous example, just because orders are received one day apart, does not mean the lead time is now one day!) It is imperative that kanban calculations consider the full LT, full DLT, and sufficient SS to cover the expected variability. The number of containers required will depend on the calculated OP and the LS negotiated with the supplier. Figure 7-6 also shows the calculation for number of cards as OP divided by the quantity per card, with any fractional values being rounded up to the nearest integer. Again, various alternative order quantities may be negotiated with the supplier, achieving the desired order point with minimal rounding required.

Other Supply/Demand Issues

Earlier we examined the impact of decreased demand, but what about other supply and demand issues? What happens, and what actions are needed, if demand increases, a supplier consistently delivers early, quality is bad (parts failing incoming inspection), or bad parts are discovered on

> **SCM Team Exercise:** Discuss examples of each of the issues above from your own company's experience. How did they impact your parts? Using your earlier ABC classification, select a sample part from each ABC category. Discuss similarities and differences in how each situation above may affect the selected parts. For example, does consistent early delivery of a C-item have a different impact than on an A-item? Should your company's response be different in each case? Have each team member develop a saw tooth chart for each issue. Compare charts and discuss any differences. What other examples beyond those noted above has your company experienced? Model each using the saw tooth chart and discuss how each might be managed using kanban.

the floor? How should you handle these and other issues? Figure 7-7 presents the issues and possible actions for the team.

If this happens, follow these actions

- **Demand increases:** If demand increases, inventory will be consumed at a faster rate than the current Order Point can cover, increasing the likelihood of stock outs.

    Action: Recalculate DLT & SS. Reset Order Point. Place an order immediately if kanban inventory is below the new OP.

- **Supplier consistently delivers early:** If your supplier does this, then your average inventory is driven up. (This phenomenon is examined in the next section below).

    Action: Renegotiate Lead Time with supplier to capture the demonstrated improvement. Recalculate DLT & SS; reset OP.

- **Supplier delivers bad parts:** Usable inventory is not replenished, creating a high risk of stock out.

    Action: Immediately reorder; expedite delivery. If the quality problem is discovered during manufacturing, usable inventory drops. If the new level is below OP, reorder and expedite if required. Develop a corrective action plan with the supplier.

Other Issues: A good technique for evaluating the impact of variations in supply and/or demand is to draw the effect on inventory over time using a saw tooth diagram. This is a powerful analysis tool to aid understanding of these impacts and to help quantify those impacts. (This technique is demonstrated on the following pages as we examine the impact of **early deliveries**).

Figure 7-7. Other Supply/Demand Issues and Possible Actions

The Problem of Early Deliveries

Figure 7-8 depicts the impact of consistent early delivery on kanban inventory. Notice that when material arrives, inventory is higher than it would have been if the full lead time had been utilized. The net effect is an increase in safety stock. This increase is equal to the inventory that would have been consumed between the time of actual receipt and the planned receipt. This, of course, also raises average inventory by the same amount. And depending on payment terms, the early delivery might mean earlier payment, negatively affecting cash flow.

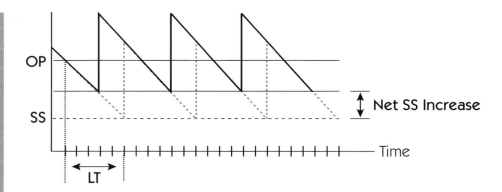

Solutions

- Deliver On Time
- Reduce LT to Actual Performance
- (Potential LS Reduction if Lead Time is Cut)

Example

- Part with 10 Day Lead Time
- Consistently delivered in 5 Days

Net Impact

- Increase in Safety Stock

Figure 7-8. Impact of Consistent Early Delivery on Kanban Inventory

Two Potential Actions for Early Delivery

The first thing the SCM team can do to preempt early delivery is renegotiate lead time with the supplier to reflect actual performance. Suppliers who have demonstrated faster delivery should be able to commit to this improved performance in the future. In these cases, the team can eliminate the negative impact of higher inventory and can actually convert it to an *inventory savings*. To do so, recalculate DLT and SS, based upon the improved lead time and reset order point. It may also be wise to

> **SCM Team Exercise:** Use your in-house supplier evaluation process to create a list of suppliers who are aligned with you regarding renegotiating lead times that reflect actual performance, those that may need help or education, and those that may be of concern. Specifically look at your critical suppliers, especially those that deliver *A*-items, and choose a few *A*-items that need an improved lead time. Then calculate SS and average inventory using both the old and new lead times. The difference is your potential *inventory savings*. Model each before and after situation using the saw tooth diagram. Referring back to Figure 7-4, are any lot size changes appropriate? What additional inventory savings will result from changing lot sizes?

renegotiate lot size if the saw tooth peaks are well above the new OP level, thereby further decreasing average inventory.

If no LT renegotiation is possible, the second alternative is to inform the supplier that early delivery is not acceptable. Educate suppliers on the negative impact of early deliveries and on the positive impact to you, their customer, of just-in-time delivery to the scheduled lead time.

SCM Team Exercise: Choose a few parts from your ABC classification and simulate a demand reduction. Use a past example or the demand patterns of a specific part. Using Figure 7-1, set the initial parameters and then perform the kanban calculation. Now plug in the demand reduction and recalculate. Finally, multiply the "excess" inventory (safety stock) times the dollar per part to determine your direct cost savings by eliminating this waste. What other costs might be avoided or reduced? (Think in terms of costs associated with storage, engineering changes, etc.)

The team so far has worked with a lot of theory. Its time to roll up your sleeves and do some real physical work.

CHAPTER EIGHT

Physical Techniques of Kanban Replenishment Systems

Up to this point, we have focused on kanban theory and concepts. The SCM team has determined ABC categories for the company's parts, has examined demand patterns and modeled them using saw tooth diagrams, and has applied statistical models to those demand patterns to determine appropriate safety stock levels. In this chapter, the team will turn its attention to the physical techniques of kanban and implementing them on the shopfloor. Before moving on, each team member should have a clear understanding of the points below.

- Key definitions of kanban: lead time, lot size, demand through lead time, safety stock, and order point. (Chapter 1)
- Definition of the Pareto Analysis (80/20 rule) and ABC classification. How to set ABC inventory categories. (Chapter 3)
- How to use the saw tooth diagram and formulas to analyze the inventory behavior of a part number. (Chapter 4).
- How to calculate demand through lead time (DLT), safety stock, and average inventory. (Chapters 4 and 7)
- How to analyze supply and demand patterns using the saw tooth diagram. (Chapter 7)
- How to determine the effect of variability on inventory and how or what to protect with safety stock. (Chapter 6)
- Single-card vs. multi-card kanban systems and the relative effect each has on average inventory. (Chapter 7).
- How to calculate number of cards needed in a multi-card system. (Chapter 7)

> **SCM Team Discussion:** Take a moment to review the bullet points above to make certain there is a clear understanding among individual team members of each point. Can everyone provide a definition for terms and/or synopsis of formulas and processes? If not, review team exercises in the referenced chapters.

Using Kanban's Manual-Visual Control Techniques

It bears repeating that kanban is a manual-visual control technique and, when properly arranged, constantly communicates the status of each part to materials/SCM people on the shopfloor. The visual controls in a well-implemented kanban system provide clear, immediate signals for many conditions, including:

- Are parts in the right location?
- Is the part on a single-card or multi-card kanban?
- Is it time to place an order?
- Has an order been placed early?
- Should an order already have been placed?

If kanban is properly set up on the shopfloor and in the stockroom, a quick walk-through should immediately make apparent any issues in these areas. Some examples:

1. Part number labels on racks must match the part number on the kanban card. (Of course, the kanban card must accurately reflect the actual material). The part number on both the kanban card and rack label must be large enough so people can easily see them from a distance.
2. Kanban cards must be visible on all multi-card stocking locations. If a card is missing, it indicates the card has been lost or pulled prematurely.
3. Kanban cards must *not* be on "work-in-process" or "go-to" locations. A card here would indicate the appropriate card was not pulled when material was moved out of a stocking location and placed in-work. (There is a slight variation on this rule for single-card kanbans, which will be made clear later).

Painted Squares on the Floor

Perhaps the simplest and most recognizable kanban technique uses squares painted on the floor between operations in an assembly process (see Figure 8-1).

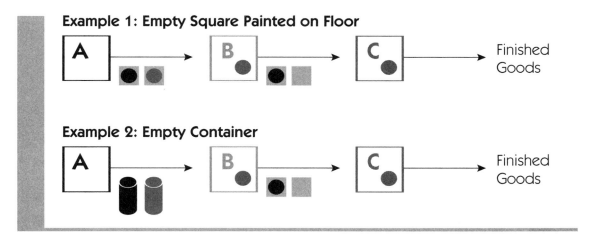

Figure 8-1. Painted Squares on a Floor

In Example 1, a unit of a gray product is ordered, triggering the pull process. Operation C pulls the corresponding gray subassembly from the designated kanban square between B and C. The empty square is the signal authorizing Operation B to produce a replacement. B then pulls one unit of the gray subassembly from the square between A and B and begins work. The empty square is the signal authorizing Operation A to produce a replacement. When each operation is complete, the material handler places the new subassembly into the square emptied by the downstream operation. In this way, product is pulled through the process only in response to actual customer demand. Shopfloor workers may only produce to fill empty kanban squares. No other production is permitted.

Example 2 is similar, except that Operation A produces black and gray units in batches, filling up containers of each, rather than producing single units. In this example, A may produce either black or gray units, in whatever quantities are needed to refill the containers. However, no production is permitted above that which is needed to fill the containers between A and B.

Technique for Storing Single Card Parts

What if there are too many parts or subassemblies required at a particular operation or work center to permit squares on the floor for each? In this case, store parts on racks or shelves within easy reach of shopfloor workers. Use the kanban calculations discussed in Chapter 7 to determine the number of kanban cards (and, by extension, the number of containers) required. *Each container should have its own kanban card.*

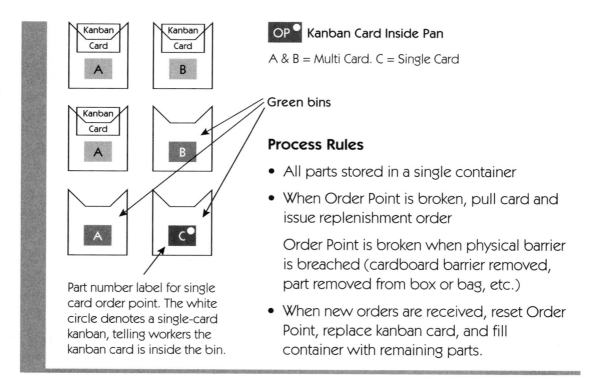

Figure 8-2. Emphasis on Parts Stored in Racks for Single Card Order Point

Figure 8-2 shows technique for storing single-card parts and multi-card parts together on shelves. For now, we will focus on part number C, which is a single-card kanban, requiring only one location. Part numbers A and B are multi-card and require multiple locations. (A and B are faded-out to deemphasize them).

This technique uses color-coded part number labels located on the front of each part bin. Green represents the active bin—the bin from which parts must be pulled. You can think of green-labeled bins as the work-in-process or go-to bins. In the case of single-card parts, you need only one location, as indicated by the green label with the white dot. Again, green indicates this is the go-to bin, and the white dot signifies that this is a single-card part. Figure 8-2 summarizes the processing rules for the single-card case. The key point concerns when to order. We will discuss managing the details of a single-card bin later in this Chapter.

Technique for Storing Multi-card Parts

Figure 8-3 shows the technique for storing multi-card parts A and B in the example. (Part "C" has been faded-out to deemphasize it.) Based on standard order point calculations (demand through lead time plus safety stock—DLT + SS), it has been determined that the OP for A requires two containers; B requires one. Adding one location as the work-in-process or go-to location gives us the arrangement in Figure 8-3. To picture this on a saw tooth chart, the two stock bins combined make up the order point, with the go-to bin holding any stock above that level. In reality, the parts are used in the opposite order, but the *function* of the two stock bins is to cover demand through lead time plus safety stock.

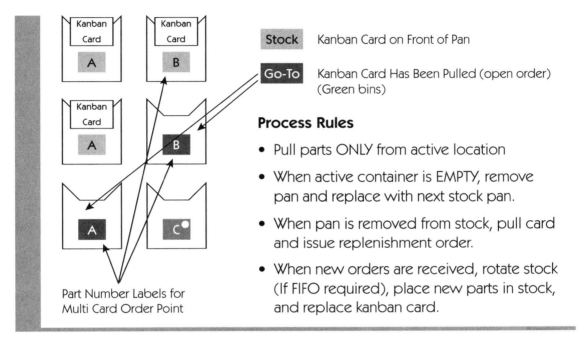

Figure 8-3. Emphasis on Parts Stored in Racks for Multi Card Order Point

A kanban card is required on each of the *stocking* locations. Parts for production are pulled only from the go-to locations. When a go-to location becomes empty, the container from the nearest stock location is moved into the go-to location, and its kanban card is pulled. The card then triggers a replenishment order. When the order arrives, the stock location is replenished and the kanban card is placed with the parts back in its proper place.

Techniques to Identify the Order Point Inside the Bin

Returning to the single-location, single-bin method, some technique is needed to clearly identify the order point level inside the bin. A good practice is to create a barrier, separating the order point parts from the remainder. This barrier must be breached in order to get at the parts comprising the order point. The act of breaking the barrier is a powerful signal that it is time to pull the kanban card and place a replenishment order.

Figure 8-4 depicts one simple technique (Option 1) where demand for a part is stable. Draw a line inside the pan at the level required to hold the order point number of parts. Then fill the bin to this level (no need to count parts, which makes this technique very efficient), place the kanban card on top of the parts, and put the cardboard divider in place. All remaining parts are placed on top of this divider.

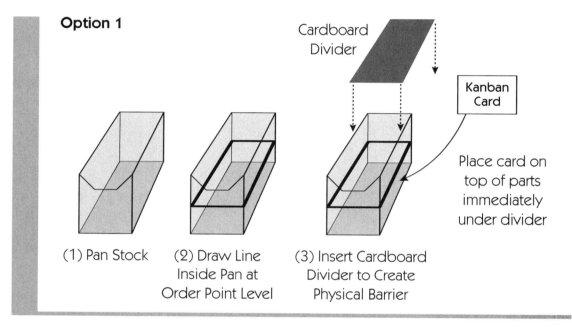

Option 1

Cardboard Divider

Kanban Card

Place card on top of parts immediately under divider

(1) Pan Stock

(2) Draw Line Inside Pan at Order Point Level

(3) Insert Cardboard Divider to Create Physical Barrier

Figure 8-4. Technique to Identify the Order Point Inside the Bin—Option 1

As parts are used, it eventually becomes necessary to breach the barrier, at which time the shopfloor worker pulls the kanban card and places a replenishment order. Parts continue to be pulled from the bin until the order is received. At this point, you fill the bin back up to the line, replace the kanban card and barrier, and put the remaining parts on top. (A first-in first-out, or FIFO, process would of course require rotating material, with new parts going in first and older parts going on top).

Another technique (Option 2) is to place the order point quantity in a smaller box or bag inside of the bin (see Figure 8-5). The shopfloor worker either attaches the kanban card to the smaller container or places it inside. All other parts are placed loosely inside the bin.

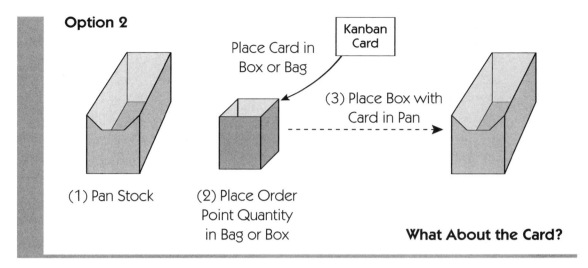

Figure 8-5. Technique to Identify the Order Point Inside the Bin—Option 2

The same processing rules apply as before. As soon as the physical barrier is breached, in this case opening the small box or bag, the shopfloor worker must pull the kanban card, triggering a replenishment order. Ideally, the box or bag is the right size to hold just the right quantity (the order point quantity), to avoid the need to count parts.

In the event that parts do not fit easily into bins or other containers, the kanban card should be captured and physically placed with the order point quantity, with remaining parts stored in the same location (Option 3). Figure 8-6 shows this technique for sheet material and bar stock.

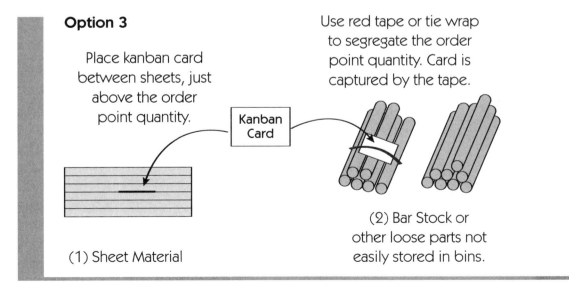

Figure 8-6. Technique to Identify the Order Point for
Parts Not Easily Stored in Bins—Option 3

- <u>Sheet material</u>. The order point in Figure 8-6 is three sheets. The kanban card is placed on top of the third sheet, sticking out slightly from the front of the sheets to make it visible. Place all other sheets on top of the card. As soon as the card is exposed (when the sheet immediately above it is pulled), the shopfloor worker pulls the card and issues a replenishment order. If possible, physically attach the card to the sheet with tape or some other means, taking care not to damage the material.

- <u>Bar stock or other loose parts</u>. The order point in Figure 8-6 is six bars. These parts are banded together with red tape, a tie wrap, or some other standard method, and the kanban card is captured by the banding. Place all other parts loosely in the same storage location. When all loose parts have been pulled, and the shopfloor worker has to cut the banding in order to use one of the captured parts, the kanban card is pulled, triggering a replenishment order.

> **SCM Team Exercise:** Using your ABC classification, choose some parts of various sizes, shapes, and quantities. Discuss possible stocking options considering the following points for each part number:
>
> a. Will all parts fit in a single bin? If not, how many kanban cards and locations do you need?
>
> b. Can you standardize bin sizes? (I recommend limiting bins to a few standard sizes).
>
> c. If you must store parts loose, (i.e., not in bins, boxes, or other containers), then how should they be stored? How should the kanban card be placed and captured?
>
> d. Is space available at the point-of-use to store all parts? If not, how much can be stored there? (We will discuss this subject later. A move card is suggested to authorize movement of parts out of the kanban locations to the shop floor in smaller quantities).

Information on the Kanban Card

The kanban card contains the minimal information necessary to inform materials people of the key points (as seen in Figure 1-2).

Part number and description: Use the P/N bar code if your company uses that technology in the operation. Always display the part number as text, using a large enough font to so people can see it from a distance.

- **Notes:** List relevant notes, as deemed appropriate.
- **Stocking Location:** Location of the part to aid in putting material in its proper place.
- **Other Information:** ABC code, lead time, lot size, or other information which may help materials personnel determine if a change in order point, number of locations, or other changes in the physical system are needed.

Handling the Kanban Card

Once the kanban card has been pulled, you will need some standardized method to ensure that the card triggers a replenishment order. I recommend using either a board or mailbox, with a clear "andon" signal to indicate that cards are ready to be picked up. In the case of a mailbox, the andon is the flag. Whenever cards are in the mailbox, the flag should be up. Periodically during the day, perhaps on a two-hour cycle, a responsible person will retrieve cards from the box and place a replenishment order (purchase order or manufacturing order) for each.

If you use permanent cards, you need some method of holding a card while the replenishment order for that part is open. When parts are received, the shopfloor worker matches the card up with its new parts and replaces it in the kanban storage bins as described in Figures 8-2 through 8-6.

A better technique, rather than using permanent cards, is to print a new card whenever a replenishment order is received. Many systems generate a receiving ticket when each order is received. This ticket, with minor modifications, may serve well as a kanban card. In this scenario, kanban cards are discarded as soon as they have been pulled (based on the previous rules for when to pull a card) and a replenishment order has been placed. Use your system's open-order-status reporting to monitor open kanban orders. In particular, I strongly recommend a management-by-exception report of past-due orders.

Options for Designating Permanent Locations

Ideally, each kanban part will have a designated, permanent location. It is essential to devise some means of clearly identifying and easily locating each part's location. One method is to divide the facility into a grid, with rows and columns (see Figure 8-7). Most facilities have support beams, or other physical structures in place, which can readily serve as "road signs." This technique easily guides people to a small geographical area within the facility.

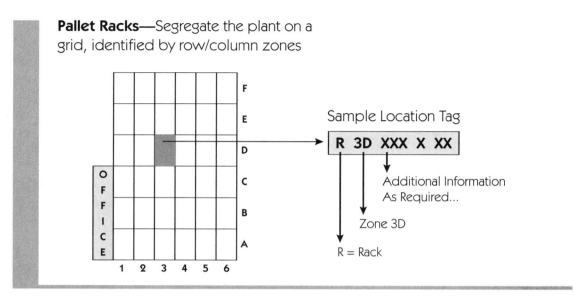

Figure 8-7. Dividing the Facility Into a Grid With Rows and Columns

Once there, the materials person will need more precise information, such as type of storage (rack, shelf, floor, carousel, etc.), and specific location within a specific storage medium. Figure 8-8 shows a coding scheme combining these elements with a storage rack with recommended location labeling.

The rack contains three vertical sections, each with multiple shelves. The rack itself is labeled as Rack A1, designating that it is a rack in location A1 of the facility. The left-most vertical section is designated section 010, the middle section 020, and the right-most section 030. Floor-level storage within each vertical section is labeled A, the next level up is labeled B, and so forth. Applying the full coding scheme, the example location in Figure 8-8 would be "R A1 020 C."

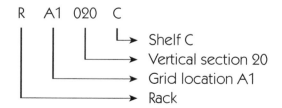

There is no need to overlabel each discreet location on the rack. Placing a sign stating "Rack A1" on the side of the rack, the numbers 010, 020,

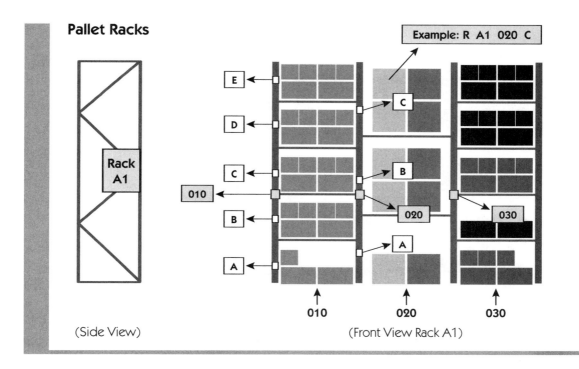

Figure 8-8. Coding Scheme, Combining Rack, Shelf, Floor, Carousel, etc.

and 030 on the vertical supports, and the letters A, B, C, etc. on the horizontal beams fully defines each location. By using the part number labeling discussed earlier, workers can easily find each part at the correct storage location.

Figure 8-9 shows another option for location coding using an assembly or manufacturing group number in place of the grid-coding scheme discussed earlier. In this option, a coil winding and assembly group might have multiple storage media in place, such as floor, rack, and shelf storage.

Figure 8-9 shows two example tags. The first example tag depicts spools of wire staged at a coil winder: F 123 A 04

The second example tag depicts parts stored in bins with a set of shelves: S 123 G 020 B

Figure 8-9 also depicts a kanban move card that can be used to signal the need for more material to be brought to the floor location. Parts are supplied from the primary storage location where the kanban reorder level (OP) and kanban order card are maintained. You need some method of ensuring quick replenishment of move cards. Suggestions include (1) placing move cards on a board that material handlers check

Pan / Floor Stock

- Use "Group" in place of "Zone"
- Include Shelf/Floor location Ids.
- Parts should be located near the point of use.
- Use Move Card to replenish from Kanban if Floor Stock is a Secondary Location.

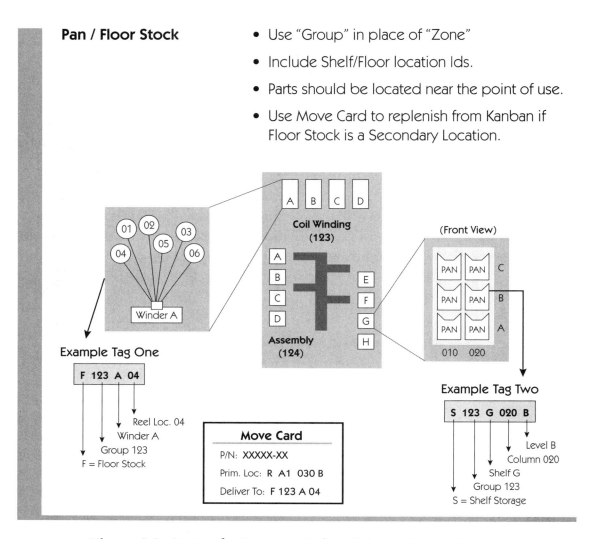

Figure 8-9. Option for Location Coding Using a Group Number

SCM Team Exercise: Survey the site and determine what option would be best suited to designate permanent locations for your parts. Do you use a grid with row and column zones to identify geographic locations? Are department or group numbers well known and clearly marked with overhead signs? If so, group number may be preferable. What color-coding standards for part number labels and kanban cards are best suited to your operation? (Color-coding is highly recommended, since colors readily communicate information to the human eye). How much detail do you need on your kanban card? Do you have the capability to print receiving tickets on incoming materials? Can you use those tickets on kanban cards? If not, what capabilities exist or need to be developed to print permanent kanban cards?

on a frequent basis and (2) using andon lights or some other highly visible signal to indicate a move card has been pulled. (Note: a typical "andon" signal is a three-color stack light. Green might indicate no cards are on the board, yellow that cards have been pulled and are on the board, and red that move cards are urgently needed to avoid a line shutdown).

Kanban Management for Raw Materials

A good working definition for raw materials is "materials that are transformed through a manufacturing or fabrication process into finished parts." Figure 8-10 shows different types of raw materials and manufacturing and fabrication processes that convert these materials into finished parts.

What Are Raw Materials?

- Materials that are transformed through a manufacturing / fabrication process into a finished part.

Raw Material Examples:

- Wood, plastic pellets, or plastic molding powder
- Metals in the form of sheets, rolls, bars, and ingots
- Fluids (colorants, release agents, chemicals)

Manufacturing / Fabrication Processes that Convert these Materials into finished parts :

- Plastic molding (thermoplastic and thermoset)
- Metal stamping, forming, die casting, milling
- Shaping operations, cutting, turning, and lathe shaping

Figure 8-10. Examples of Raw Materials and Manufacturing and Fabrication Processes

Challenges and Recommendations for Managing Raw Material

Because of the nature of raw materials, they present some unique challenges for effective kanban control. Figure 8-11 shows some of the challenges inherent in managing raw materials as well as some techniques for effective kanban control.

Challenges

- Materials are not easily counted.
  . . . Must be weighed or measured.

- Portion of raw material may be returned to stock upon completion of fabrication operation.

- Materials may not arrive in precise order quantities.

What Can You Do?

Recommendations

- If possible, work with supplier to subdivide into "units" compatible with a multiple-card kanban approach.

- Set up kanban to allow order point to be easily (visually) obvious (e.g., material moved from Stock location to WIP).

Otherwise ...

- Establish a ledger system

Figure 8-11. Challenges and Recommendations for Managing Raw Material

Let's look at some of the challenges associated with raw material kanban.

- **Materials are not easily counted.** For example, with a coil of steel weighing one ton, and a reorder point of 500 pounds, it is difficult to determine visually how much material remains and when you have reached the reorder point. Weighing the coil and recording its weight prior to placing it in the rack may be required.

- **Manufacturing operations sometimes require portions of raw material to be returned to stores after completion of a particular order.** Continuing the example of a coil of steel, you may have to return the one-ton coil to its storage location after stamping some quantity of parts. Weighing the coil before returning it to the rack will indicate whether or not you have reached the reorder point. If so, pull the kanban card and place a replenishment order.

- **Materials may not arrive in precise order quantities.** Weighing incoming material can help ensure that, in fact, you have received the correct amount of material. In addition, small variations in material dimensions or fill levels in containers can dramatically affect the actual quantity of parts that a manufacturing process will yield.

Now let's look at a few options for establishing good kanban controls. One option is to work with the supplier to subdivide materials into units that are compatible with manufacturing lot sizing to minimize the need to return materials to stores once they've been pulled and to enable the use of the multiple-card kanban approach described earlier. This combination allows kanban cards to be pulled at the time material is removed from the kanban rack, in keeping with normal kanban rules. In addition, the multi-card approach minimizes the inventory investment required on each material, enabling visual management techniques rather than requiring you to measure or weigh material multiple times, which is costly.

Employing a Ledger System for Managing Raw Materials

If the recommendations discussed in Figure 8-11 prove unpractical, another option is to employ a ledger system using a simple 3x5 card (see Figure 8-12). This ledger can manage materials that you must weigh or measure to determine how much material is on hand and when you have reached the order point.

Each raw material part number has its own card, indicating some basic information such as part number, order quantity, unit of measure, and order point. As material is received, the shopfloor worker records a new balance. After each use of the material, the shopfloor worker compares the remaining balance to the order point. If the remaining balance is less than order point, the shopfloor worker pulls the kanban card and places a replenishment order. You can store the kanban cards in a convenient, easily accessible place rather than with the material; however, you should still use normal rack labeling (location code and part number) to enable workers to readily locate and identify material.

Cautions and Recommendations for Setting the Order Point

When determining the order point for raw materials, a few final cautions and recommendations should be considered (see Figure 8-13).

Let's take a closer look at these cautions and recommendations.

- **Raw material can be used to make many different parts.**
  The SCM Team must accurately document the expected amount of

Part Number: XYZ - 123 - A
Order Qty: 1000
U/M: lbs.
Order Point: 750
Loc: R xxx

Date	Qty. Received	Qty. Consumed	Balance	PO #	Due Date
			1,000 Lbs.		
1/2		200	800 Lbs.		
1/5		250	550 Lbs.	12345	1/10
1/9		125	425 Lbs.		
1/10	1000		1,425 Lbs.		
1/15		200	1,225 Lbs.		

Key Points

- 3x5 Index Card
- Keep Kanban Card Readily Accessible
- Pull Card When Balance < OP. (Ref. PO at 1/5)

Figure 8-12. Ledger System for Managing Raw Materials

Cautions

- One Raw Material can be used to make multiple parts.
- Each make-part has its own Lot Size.
- Offal can be different depending on make-part.

What Can You Do?

Recommendations

- When calculating OP, make sure OP > Raw Material Required for the parent part with the largest material consumption.
- Consider offal when structuring BOMs. May need to work with Accounting to properly account for costs.

Figure 8-13. Cautions and Recommendations for Setting the Order Point

material consumed in the production of each part (usually in the part's bill of material), so that they record accurate material usage data. You may need to work with the accounting department to track costs properly.

- **Each make-part will likely have its own Lot Size.** This is of special concern. A good rule to follow is to set a lower limit on the raw material order point equal to the amount of that material needed to produce one lot size of the part having the highest material consumption. That is, when calculating OP, make sure OP > raw material required for the parent part with the largest material consumption. In the example in Table 8-1, the order point of 362 lbs. (based on average DLT plus safety stock) is not enough to cover one lot size of part C, which requires 500 lbs. If an order arrives for part C with inventory remaining above 362 lbs. but below 500 lbs., a stock out will occur.

- **Offal (waste material left over from a manufacturing process) can be different depending on make-part.** Consider offal when structuring BOMs to ensure that you properly record the total amount of raw material needed to produce each part. For example, in a plastic molding operation, some raw material is wasted at the beginning and end of each run. Or, when stamping multiple parts from a sheet of steel, some steel remains between parts. You must consider the total amount of raw material consumed, including offal, when calculating kanban order points and lot sizing rules. In an assembly operation, the equivalent of offal is called "shrinkage." If some fall-out is expected during the assembly process, you must base kanban order points and lot sizes on the total amount of inventory consumed, including shrinkage. For example, if an assembly operation typically produces 5 percent scrap, and enough material is introduced at the beginning of the line to produce 100 pieces, only 95 good products will come off the end of the line. Kanbans for the component parts and raw materials must consider the full 100 as having been consumed.

Table 8-1. Calculating OP to ensure OP > Raw Material Required
for the Parent Part with the Highest Material Consumption

P/N	Lot Size	RM per piece (lbs)	Raw Mat'l per Order (lbs)	No. Orders per Month	Annual RM Usage (lbs)
A	50	1	50	2	100
B	20	2	40	1	40
C	500	1	500	2	1,000
D	25	3	75	1	75
					1,215

If the lead time is 5 days for each part, the Avg. DLT = (1,815 / 21) x 5 = 289 lbs.

Adding 25% Safety Stock would give an Order Point = 362 lbs.

SCM Team Exercise: Discuss any raw materials used in operations. Does a fabrication shop exist in your plant? What raw materials are used? Are multiple finished parts produced from a common raw material? If so, review the raw material issues above, especially the issue of make-part lot sizing. Will calculating a kanban OP based on average demand cover the material required to produce the largest lot size of the parent make-part?

Next, review the suppliers' lot sizes on raw materials. Will these quantities and packaging techniques accommodate how you would like to set up kanban? What adjustments might be needed?

SCM Team Exercise: Is shrinkage an issue in your assembly lines? If so, are shrink factors included in your bills of material? If not, determine how you will compensate for this when calculating kanban order points.

CHAPTER NINE

Kanban Maintenance

Maintenance of a kanban system is critical to its ongoing success. The SCM Team must routinely scrutinize the system to ensure that it is meeting the company's inventory and service objectives. A process of ongoing evaluation, including such things as routine audits and benchmarking, will ensure that the team identifies opportunities for improvement, enabling them to drive continuous improvement of the system. Likewise, the team must evaluate individual parts on a regular basis to ensure that they are set up properly and that order points, safety stock, and other kanban parameters are adequate to cover current and anticipated demand levels. Figure 9-1 summarizes the five basic principles for maintaining the kanban system that we will discuss in this chapter.

Kanban Adjustment Report and Card Maintenance

A key tool for maintaining kanban parameters is a Kanban Adjustment Report (see Figure 9-2). The primary purpose of this report is to identify parts whose order points are either too large or too small given anticipated demand levels. The report first calculates required order points (DLT + SS), then compares these values to the current order points as stored in the company's database. Ideally, this database will be an integral part of the company's ERP or appropriate materials planning system; however, in smaller companies, or in pilot implementation phases, it may be a spreadsheet or desktop database. A responsible analyst on the SCM Team should evaluate reported values that differ significantly from current levels.

1. **Kanban Adjustment Report (Current OP is Too Large or Too Small)**

   ✔ Parameters (Usage, SS, LT, LS, etc.)

   ✔ Filters (Too Large / Too Small Rules)

   ✔ Frequency (By ABC)

2. **Card Maintenance (This should always be up to date)**

   ✔ Part Identification

   ✔ Supplier

   ✔ Location, etc.

3. **Stock out / Expedite Tracking**

   ✔ Good indication of where kanban is weak

   ✔ Data issues?

   ✔ Execution Issues? (Internal or Supplier)

4. **Supplier Performance**

   ✔ Delivery Performance

   ✔ Quality Performance

   ✔ Impacts on Safety Stock

5. **Kanban Audits**

   ✔ Based on standard kanban policies and procedures

   ✔ Drive Standardization

   ✔ Goal of Continuous Improvement

> **Remember:**
> Maintaining the kanban system is a key to its success.

Figure 9-1. Five Basic Principles for Maintaining the Kanban System

The SCM Team must determine what constitutes a "significant" change, as well as define specific data filters for the report and determine how frequently to review the report. A general rule might be to evaluate *A*-items frequently, perhaps weekly or monthly, since those items will likely have the smallest proportion of safety stock, and therefore will be susceptible to stock out, given relatively small increases in demand. Conversely, small decreases in demand on *A*-items may represent significant inventory reduction opportunity because these items represent 80 percent of the annual inventory investment. *C*-items will likely have much higher safety stock levels and are therefore less susceptible to minor variations. The report might also identify parts whose lead time, lot size, or other parameters differ from company objectives.

Parts with Order Point Too Large. (Need to adjust OP down).

Part Number	ABC	Primary Location	Avg. Daily Usage	Lead Time	DLT	SS%	OP	Lot Size	No. Cards Reqd.	Current No. Cards	No. Cards to be Removed	Std. Cost	OP Reduction ($s)
ABC	B	R A5 030 B	95	10	950	50%	1,425	500	3	4	1	$0.25	$125
DEF	C	R B4 060 D	2,375	15	35,625	75%	62,344	25,000	3	5	2	$0.05	$2,500
GHI	A	S B6 020 A	805	5	4,025	25%	5,031	2,500	3	4	1	$0.75	$1,875

Parts with Order Point Too Small. (Need to adjust OP up).

Part Number	ABC	Primary Location	Avg. Daily Usage	Lead Time	DLT	SS%	OP	Lot Size	No. Cards Reqd.	Current No. Cards	No. Cards to be Removed	Std. Cost	OP Increase (%)
LMN	A	R B3 020 B	58	10	580	25%	725	500	2	1	1	$2.50	100%
OPQ	C	R B4 050 D	225	8	1,800	75%	3,150	1,000	4	3	1	$0.02	33%
RST	A	S A2 030 B	510	5	2,550	25%	3,188	1,200	3	2	1	$0.75	50%

Notes:

- Too-large criteria: OP Reduction > $100. (Inventory reduction must be worth the cost of making the adjustment).

- Too-small criteria: OP Increase > X%. (Only adjust OP up if the increased demand has eroded too much of your safety stock. On Part "OPQ," an increase of 33% will consume a large portion of the desired 75% safety stock. To ensure a high level of availability on that part, its OP should be increased).

- DLT, OP, No. Cards Reqd., No. Cards to be Removed, and OP Reduction are **calculated** values.

Figure 9-2. Kanban Adjustment Report for Maintaining Kanban Parameters

99

Other exception reports may include the following:

- Part numbers with safety stock out of expected range: compare each part's safety stock to an expected range (sigma level, percent, cost, etc.). Report only parts outside of the range.

- Parts with significant difference between average monthly receipts (based on PO or manufacturing order receipts) and average monthly usage (based on inventory issue transactions or backflush). For *A*-items these two values should be very close to each other, perhaps within 10 percent, because they tend to have high-volume repetitive usage, and should be ordered frequently. *C*-items, on the other hand, may have very erratic usage and should be ordered infrequently. If the company has just recently received a PO, the average monthly receipts could be quite a bit higher than the average monthly usage. A reasonable range for *C*-items might be 50 percent. *B*-items might be expected to be within 25 percent. Parts outside of these ranges may have bill of material (BOM) errors, shrinkage, scrap or offal rates which have not been adequately considered, or supplier quantity issues (i.e., different actual quantities in packages than what the system says). If this is the case, initiate root-cause analysis and corrective action on exceptions to ensure that these issues are addressed.

SCM Team Exercise: Discuss options for creating a Kanban Adjustment report, and the exception reports listed above. How might the SCM Team create these reports? What other reports, or corrective action processes, might be applicable for your environment?

The SCM team must keep the kanban cards up to date as well. When data changes from the values printed on the cards, produce new cards to replace the out-of-date cards currently in use. Examples include changes in stocking location, card quantity and supplier. One way to ensure a regular review of permanent kanban cards is to use a two-part, or perforated, card. When new parts are received and put in stock, the material handler verifies that the information on the card is correct. If the card needs corrections, the material handler notes them on the removable portion of the card and gives it to the person responsible for card maintenance. However, the best technique, as mentioned before, is to use the receiving ticket as the kanban card or to print an additional document at the same time that can serve as the kanban card. Doing so ensures that cards are always up to date.

Supplier Performance Reports and Stock out/Expedite Tracking Report

Two other tools for monitoring the effectiveness of the kanban system and for keeping it synchronized with current conditions are the stock out and expedite tracking report and the supplier performance reports.

Stock out and Expedite Tracking

If the company's computer systems record each occurrence of stock out or expediting, you can automate this report. Otherwise, train the people responsible for expediting to keep a manual record of these occurrences. A good technique is to report the number of stock out and expedite occurrences by product line by month, with a rolling 12-month window. This allows you to monitor the overall level of stock outs and expedites, and to see trends. Evaluate specific part numbers that exceed a threshold of occurrences, for example, two or more occurrences in a single month, or three occurrences in a three-month period. Parts with a history of stocking out or of frequent expediting may be set up improperly. Evaluate kanban parameters such as lead time, lot size, order point, and safety stock to determine if there are any obvious data issues at fault. It may also be possible that the execution side has broken down. Potential execution issues include lost kanban cards, cards not being pulled on time, or deviation from stocking or location standards.

Supplier Performance

It might also be possible that supplier issues exist. A supplier performance report, or reports, can be useful in identifying such issues. Among the points to monitor are poor on-time delivery, partial deliveries, incorrect delivery quantities, and quality issues. In the short term, you can compensate for such issues with additional safety stock; however, you should pursue corrective action aggressively to avoid unnecessary inventory investment and to ensure a high level of parts availability for manufacturing.

Kanban Audits

Kanban audits can be an effective tool to ensure the viability of the kanban system. Routine audits that cover all related processes can ensure they are running optimally. This covers such activities as:

- Order points and safety stock in line with standards
- Parts set up per physical standards
- Kanban cards pulled on time
- Location codes per standard and physical locations match system records
- Information on kanban cards is up to date
- Kanban cards in place per standards and properly positioned
- Housekeeping (kanban locations are kept neat and orderly)

The goals of a kanban audit are to:

1. Ensure the proper functioning of the kanban system today.
2. Identify opportunities for continuous improvement.

For this reason, it is a good idea to share the kanban audit results with all functions and encourage the responsible parties to take immediate action to correct problems. Repetitive problems may indicate a systemic breakdown with your replenishment system and warrant further action by the SCM team or by management to identify and correct the root causes.

> **Lean Note:** When establishing the physical elements of kanban such as racking, labeling, standardizing bins, and creating kanban cards, 5S is a powerful weapon to ensure each area is set up with excellence (see Figure 1-12). *Sort* out anything that doesn't belong in or near kanban storage. *Scrub & shine* the entire area to achieve the "wow" effect. Using 5S to establish a clean and orderly kanban structure will dramatically improve kanban's operation and appearance. But the real power of 5S is to *standardize* and *sustain* the efforts of the first three Ss. Kanban audits can help drive and establish these two Ss throughout the organization.

SCM Team Exercise: Review the current way you maintain your replenishment system. Determine what existing practices can continue to add value as you implement kanban. Can you make some minor changes to existing reports or audits to incorporate the points discussed in this chapter? Determine what organizational changes you may need. Can you incorporate the activities discussed in this chapter, and in your own discussions through the earlier team exercise, into existing positions?

On the supplier side, determine if existing supplier performance reports exist. If not, how can you create them? What steps will the SCM team, or purchasing management, have to take to ensure that you address issues with suppliers in a timely manner? (Review Chapter 4 and revisit the points discussed there concerning appropriate responses and actions for supplier issues). What proactive communication should purchasing management have with suppliers to ensure they are on-board with anticipated changes?

CHAPTER TEN

Kanban Implementation

This chapter outlines an approach for managing the implementation of a kanban system in a manufacturing plant. There are two fundamentals to stress:

1. Clear roles and responsibilities
2. The importance of managing the implementation as a formal project

Roles and Responsibilities in Kanban Implementation

In any business process, a key to success is having clearly defined roles and responsibilities. Virtually every function in the plant has a role in the kanban system. Each person must understand his or her role, and management must ensure that the individual elements support the overall process. Figure 10-1 identifies some of the key participants in the transition to kanban.

Let's take a closer look at some of the more critical participants in the project and how each contributes to the ongoing success of your company's transition to kanban. A brief description of relevant roles and responsibilities follows:

- **Management:** As in any major change effort, management buy-in and support are critical. Supervisors must ensure they understand the kanban system and their respective departments' roles; they must also monitor and control day-to-day activities.

Everyone Has a Role

- Management, Supervisors

- Project Leader / SCM Team

- Line Leaders, Hourly Workers, Assemblers, Shop Floor Workers, etc.

- Receiving, Purchasing (Buyers), Material Handlers

- Material/Kanban Analysts

- Manufacturing Engineers

- Finance Personnel

Figure 10-1. Key Participants in the Transition to Kanban and its Success Over Time

- **Project Leader/SCM Team:** The core team members must become experts in the new process. They will provide management with education and training materials (draw from this workbook and/or other resources to develop appropriate education and training programs), lay out a clear implementation plan, and manage the project to the plan. (A high-level plan is presented later in this chapter. You may add detailed activities as necessary to address specific challenges in your company).

- **Line leaders:** Kanban execution, as stated previously, is a manual-visual process. Hourly workers (shopfloor workers, assemblers) can make or break the system, and it often falls to line leaders to ensure that each individual is following kanban disciplines on the shopfloor. They are responsible for ensuring that material is located properly and that kanban levels are reset per the standards. Line leaders make certain that kanban cards are clearly visible, placed in the correct locations, and pulled at the appropriate times. It is also their responsibility to ensure that hourly workers remove material only from the appropriate go-to locations and containers.

- **Material Handlers/Receiving:** These departments manage material coming into the kanban system, whether from internal or external suppliers per the kanban standards. The receiving department matches incoming orders with kanban cards and ensures that the quantity received matches the order quantity. Material handlers bring materials from receiving to kanban stores, ensure that FIFO rules are followed, reset kanban order points, and place kanban cards with material according to the standards.

- **Buyers:** Purchasing's role is critical in managing the incoming supply chain and driving supplier improvements in such areas as on-time delivery, quality, price, use of standard containers, and speed and frequency of delivery. (Refer to Chapter 5 for lead time and lot size guidelines to help drive the speed and frequency objectives).

- **Material Analysts:** In an MRP environment, MRP analysts play a key role in ensuring the material requirements plan is valid and is executed in a timely manner. Likewise, in a kanban environment, kanban analysts ensure that order point levels are sufficient to cover expected demand. They work with material handlers or other assigned shopfloor workers to set up the appropriate kanban technique as needed. Kanban analysts also conduct routine audits to ensure that workers on the shopfloor are properly maintaining kanban on the shopfloor and in the stockroom. Unlike MRP analysts, kanban analysts do *not* execute the kanban pulls. Shopfloor workers pull cards only when indicated by the manual-visual triggers. Analysts ensure the system is set up and maintained properly to enable the shopfloor execution processes to function as designed. These analysts also drive overall kanban material levels to meet management's inventory objectives, perform periodic ABC updates, and coordinate between manufacturing and materials/purchasing to ensure that the objectives of each (which are often at odds with each other) are met and balanced.

- **Manufacturing Engineers:** As discussed in the next section, part of the kanban implementation plan is to create a layout on the manufacturing floor that supports the kanban processes while enabling smooth material flow into and through manufacturing and assembly processes. Manufacturing engineers are typically responsible for the factory layout and must understand the kanban process and *be included as members of the implementation team.*

- **Finance:** As with any transformation project, there are costs associated with implementing kanban. The SCM team and management will need to work with finance to decide how to budget for these costs (i.e., which costs should be capitalized, made indirect, base-cost, etc.). Other activities requiring support from finance may include creating a post-deduct (backflush) capability if one does not already exist, incorporating expected inventory value changes in the financial or operating plan, and participating in any other activities affecting the financial statements of the company.

To ensure that everyone in the organization has a voice, the ideal make-up of the SCM team should consist of representatives from across the organization. The ideal team leader, selected by management, has overall project responsibility and works with a management sponsor. Once the team has developed a clearly defined kanban process, team members will need to suggest job description and/or organizational changes to management. The following points represent a general approach to doing this.

- **Clearly Define Everyone's Role.** Use the previously described descriptions as a starting point and modify them to fit the organizational structure and culture of your company.

- **Perform Roles and Responsibilities (R&R) Gap Analysis—Adjust as Required.** (Gap analysis describes the gap that exists between new R&Rs and those that exist to support the current process). Management has the lead role in creating and/or modifying job descriptions to align with the new processes and reorganizing as necessary during the transition. Figure 10-2 shows a typical gap analysis on the new "Kanban Analyst" position.

New Roles and Responsibilities	Currently Performed By	Gap? If Yes, Describe	Action Plan	Barrier (s)	Assignment
Perform semi-annual ABC Update	MRP Analyst	No			
Weekly Analysis of Order Points (Too Large / Too Small)	None	No process currently exists. Need to create.	Prototype OP calculation in Excel. Develop new OP Analysis Report.	Data not readily available. (IT Resource needed).	John
Ensure parts are physically set up properly on the shop floor	Stock Room Supervisor	Rack labeling does not meet kanban standards. Housekeeping is poor (cluttered racks, misplaced materials, etc.).	Establish joint responsibility between Kanban Analyst and Stock Room Supervisor. Set up and label racks per new kanban standards.	Organizational reporting relationships of key players differ. Analyst and Supervisor should report to same manager.	Mary
Perform monthly kanban audits	None	Audit process does not exist.	Create an audit process. Assign to Kanban Analyst. Materials & Manufacturing Managers to also perform monthly audits.	Auditor Training	Bill. (SCM Team to provide training).

Figure 10-2. Typical Gap Analysis on a New "Kanban Analyst" Position

To help ensure success, I strongly recommend that the team manage the kanban implementation as a formal project. Use project management tools such as a Gantt chart, action item lists, and open issues lists to track kanban's implementation. The project leader should be well versed in these techniques and have the confidence and support of management. A high-level project plan is presented in the next section.

> **SCM Team Exercise:** Discuss the roles and responsibilities related to kanban implementation and how they correspond to your company. Make a list of the people who will fill these roles. Depending on the size of the company, one person may have two or more roles. You may have to repeat this exercise more than once. During the SCM team's education phase, it is helpful to grapple with this topic. However, as your kanban implementation plan is developed and as the team gains experience through the pilot area and early implementation areas/groups, you will likely want to revisit and fine-tune the roles and responsibilities of specific individuals or groups.

Using a Project Plan for Implementing Kanban

Having established the roles and responsibilities, it is now time for the team to begin implementing kanban. The conversion to a kanban system is a major change for any company, especially for one accustomed to using MRP. Therefore, the SCM team should manage the conversion using a project plan. Figure 10-3 shows a high-level overview of the tasks typical in a kanban implementation project. Of course, you can change the sequence and duration of tasks to suit the specific needs and culture of your company.

The top section indicates up-front activities such as selecting the SCM team, documenting kanban standards, and selecting a pilot area. This is followed by more detailed implementation steps. The pilot area should be representative of the plant's products and manufacturing processes. Ideally, it is small enough to manage and makes a good showcase. Select an area with a high probability of success and then build on that success as you implement kanban in other groups. At the end of this initial phase, I recommend a review step during which you audit the initial implementation against the written standards and make any adjustments prior to rollout in other departments. Activity number 16 in Figure 10-3 is a good place to perform this review.

The bottom section of Figure 10-3 indicates a repeating series of steps for implementing the process throughout the plant. Depending on the availability of key resources, the SCM team may implement kanban in multiple manufacturing departments in parallel, performing each activity across multiple departments. For example, you can do the kanban OP calculations for multiple groups at once then follow up by purchasing

Week Number: 1 2 5 6 7 9 10 11 12 13 14 15 16 17 18 19 22 23 25 26 27 28 29 30 31

Task #		Notes
1	Select & train the SCM team	
2	Develop and document a "Kanban Standard"	(Update as needed.)
3	Select a pilot area	
4	Select & train pilot team	
5	Validate key kanban parameters (LT, LS, etc.)	(Ongoing. Recalculate kanban levels as needed).
6	Standardize container types, sizes, and storage	(Purchase shelves, racks, containers as needed).
7	Develop kanban calculation / maintenance process	
8	Perform kanban calculations (OP, SS, #Cards)	
9	Determine space requirements	
10	Develop manufacturing / material layout	
11	Print rack labels & kanban labels	
12	Convert physical layout	
13	Update system records	
14	Train pilot group	
15	Go-live	
16	Audit, adjust, update the standard	
17	Select next group	
18	Repeat steps 4-17 (Repeating Schedule below)	
19	Enhance computer systems as needed	

Repeating Schedule — Week Number: 1 2 3 4 5 6 7 8 9 10

Task #		Notes
1	Train new team members	
2	Validate key kanban parameters (LT, LS, etc.)	(Ongoing. Recalculate kanban levels as needed).
3	Standardize container types, sizes, and storage	(Purchase shelves, racks, containers as needed).
4	Perform kanban calculations (OP, SS, #Cards)	
5	Determine space requirements	(Select next group. Implement in parallel).
6	Develop manufacturing / material layout	
7	Print rack labels & kanban labels	
8	Convert physical layout	
9	Update system records	
10	Train group	
11	Go-live	
12	Audit, adjust, update the standard	

Figure 10-3. High-Level Project Plan for Kanban Implementation

the required bins and racking for all these groups at one time. In this way, the company can accomplish the full conversion to kanban much more quickly. At the conclusion of each group's or each department's implementation, conduct an audit to ensure compliance with the standards and make any necessary adjustments. This is a time when the team can capture lessons learned and suggest improvements to the standard that can help improve the process for the entire plant.

SCM Team Exercise: Design a draft project plan. A good preliminary exercise is for the team to walk around the plant with a pencil and pad and do a simple value stream map for material replenishment. Figure 10-4 provides a definition of value stream and value stream mapping. (*Note*: A value stream map can encompass only a portion of the full value stream—in this case, material replenishment only). By mapping the process from beginning to end and drawing a visual representation of every step in the material and information flow, the team will have a current state map of the material replenishment process to work from as they determine the timeframe and challenges for kanban implementation.

In addition to the value stream map, the team should discuss possible criteria for selecting a pilot area, considering for example the following points: size of the group, number and mix of products produced, stability of customer demand for the products, number and mix of component and raw material parts, and willingness/readiness of the group to be the pilot. Once both of these exercises are complete, the team discusses each task in the project plan and decides which team member(s) will be responsible for each. It is also important to define potential obstacles or challenges you may need to overcome. An open and frank discussion with management is a good idea at this stage. Remember that management's role is to clear the path. Involving management at this stage of planning better prepares them to fulfill their role.

Value Stream

The value stream includes all value added and non-value added activities a company's products or services currently must flow through to be delivered to the customer.

Value assumes that you are creating something of value that a customer is willing to pay for. **Stream** refers to a sequential flow of activities needed to create work units and deliver them to the customer.

There are three areas of the value stream that overlap and flow together:

1. **Problem solving (administrative):** The flow from the product design to development and launch.

2. **Information management (administrative):** The flow from a customer's order to cash, including production scheduling and other non-production (office) activities for delivery of the finished product or service.

3. **Physical transformation (manufacturing):** The flow from the transformation of raw materials to a finished product into the hands of the customer.

Value Stream Mapping

Value stream mapping is a paper and pencil tool that provides a visual understanding of the flow of material and information as a product or service makes its way through the value stream—from the time it arrives as raw material (or information), through all manufacturing and/or administrative process steps to a finished and delivered product/service to the customer. Mapping is a critical step in lean conversions because it visually highlights the sources of waste and shows where to apply lean techniques to create a future state that eliminates wastes and improves the overall flow of value to the customer. A value stream map can be bounded, encompassing only a portion of the full value stream.

Modified from: *Lean Thinking* (1996) James Womack and Daniel Jones, *The Complete Lean Enterprise* (2004) Beau Keyte and Drew Locher, *Value Stream Management for the Lean Office* (2003) Don Taping and Tom Shuker, and *Learning to See* (1998) Mike Rother and John Shook.

Figure 10-4. Defining Value Stream and Value Stream Mapping

CHAPTER ELEVEN

Stabilizing Production

In the introduction, I made the point that you must stabilize production in order for the kanban processes to be the most effective. In a pure pull system, production will follow demand as it occurs, increasing and decreasing as required by the market. APICS ("Basics of Supply Chain Management," Version 2.1, August 2001) refers to this as a *chase strategy*, as depicted in Figure 11-1. While this approach minimizes finished goods inventory, it requires great flexibility in production capacity, both internally and externally with suppliers. It also requires that plant inventory (parts, components, subassemblies) be sufficient to cover the spikes in demand, or in the case of seasonal demand, that inventory be raised prior to the peak seasons and reduced during off seasons.

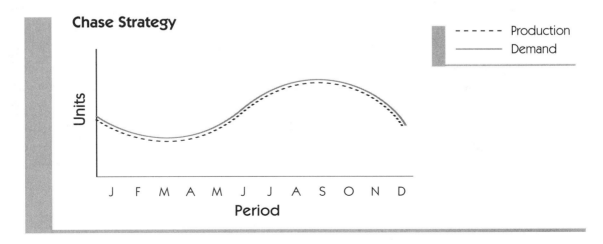

Figure 11-1. Chase Strategy for Stabilizing Production

At the other extreme is the *level production strategy*, depicted in Figure 11-2. In this approach, the company must forecast sales and plan the finished goods inventory to cover sales. This strategy holds production constant throughout the year, allowing inventory to build up during off seasons and selling off inventory during peak seasons. This approach requires a seasonal investment in finished goods inventory but has the advantages of stabilizing production, minimizing plant inventory, and stabilizing the supply chain. It enables kanban pull processes to function most effectively.

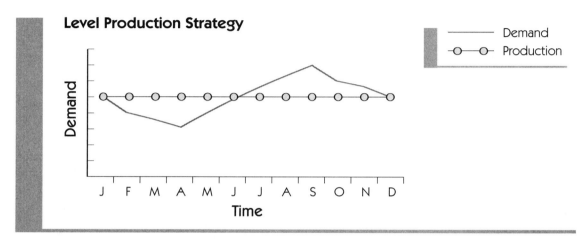

Arnold, J.R. Tony; Chapman, Stephen N., Introduction to Materials Management, 4th Edition, © 2001. Reprinted by permission of Pearson Education, Inc., Upper Saddle River, NJ.

Figure 11-2. Level Production Strategy for Stabilizing Production

A middle-ground approach, sometimes referred to as a *combination strategy*, stabilizes production for fixed periods of time, raising production prior to peak seasons and lowering it following the peaks to avoid overbuilding inventory (see Figure 11-3). Kanban can work very effectively in such a strategy; however, kanban order point levels should be raised prior to periods of high production and lowered during periods of lower production in order to balance inventory investment properly during each production period. This is especially important on *A*-items and, to a lesser extent, *B*-items. Since the inventory investment to adjust *C*-items is minimal and the work required to adjust them is extensive, you may choose to hold *C*-item order points constant throughout the year, at a level consistent with peak demand.

Regardless which strategy the SCM team selects, it is wise to work with suppliers in advance to ensure their inventory, production capacity,

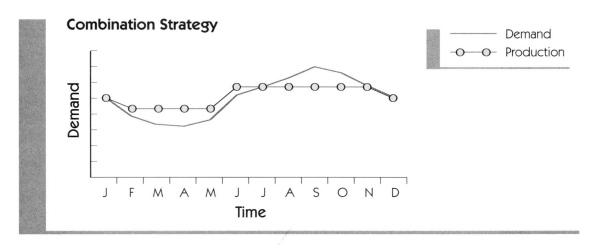

Arnold, J.R. Tony; Chapman, Stephen N., Introduction to Materials Management, 4th Edition, © 2001. Reprinted by permission of Pearson Education, Inc., Upper Saddle River, NJ.

Figure 11-3. Combined Strategy for Stabilizing Production

and logistics processes are compatible with your own company's production plans.

> **Lean Note:** Kanban works best with stable, predictable demand. Since customers do not always behave in a predictable manner, companies must often impose internal stability through level-loading, a method for developing a daily build schedule that levels the peaks and valleys of the overall quantity of the final product you build each day. Building to a level-loaded schedule also drives stability through product lines that have been balanced to takt time, and allows kanban to transmit this level schedule upstream through the manufacturing operation and back to external suppliers. To ensure that the level-loading techniques do not build excess inventory and the *muda* associated with it, the Min/Max model described in this chapter ensures that a level of stability is achieved while simultaneously maintaining reasonable inventory levels.

Creating a Hybrid Push/Pull Model—Min/Max Stabilization Process

One technique for stabilizing production of specific products is to replace traditional order point replenishment with a Min/Max model, combined with fixed daily or weekly orders, in essence creating a hybrid push/pull model. In a pure order point model, replenishment of a product is triggered only when inventory level drops below the order point. At that point, a standard fixed lot size is ordered. The timing of replenishment orders will vary as demand patterns fluctuate, creating erratic production and erratic raw material requirements in the upstream supply chain. Figure 11-4 illustrates how a Min/Max model absorbs these fluctuations, stabilizing both internal production and the upstream supply chain.

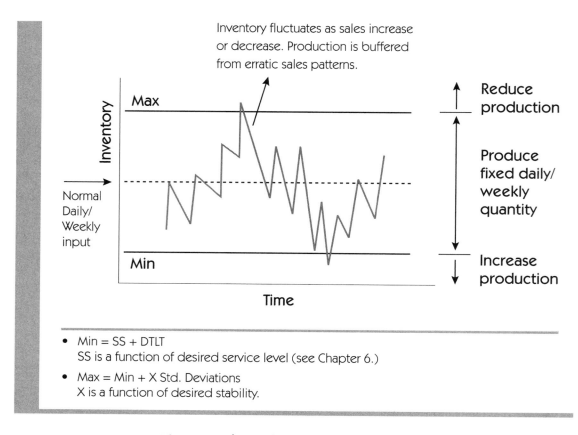

Figure 11-4. Min/Max Stabilization Model

Each day the company produces a standard daily quantity and places it into inventory. As long as inventory remains between the Min and Max, no change is made to the production schedule. Fluctuations in sales will drive inventory levels up or down, but these fluctuations have no affect on production unless they cause inventory to rise above the Max or fall below the Min. If inventory rises above the Max, production is slowed (or even halted) on that item, so that inventory quickly falls back into the range between Min and Max. Similarly, if inventory drops below the Min, production is increased for that item. Notice in Figure 11-4 that you would consider production changes on only two days—

once when inventory rose above the Max and once when it fell below the Min. On all other days, production output would remain completely stable. A variation on the exception rules might be to adjust the daily/weekly schedule only if a greater-than-Max or lower-than-Min condition existed for two or more consecutive days. The goal is to stabilize production while simultaneously ensuring high levels of customer service and maintaining reasonable inventory levels. The company can be flexible in its Min/Max scheduling rules so long as these objectives remain in balance.

The Min/Max process is product-specific and is recommended for the highest-volume products. By stabilizing production of high volume products, you achieve a high degree of overall stability without having to produce every product to a pre-defined production plan. In most cases, the Pareto rule will apply, so that applying the technique to the top 20 percent of products stabilizes production on 80 percent of sales. Use normal order point processes to manage the remaining 80 percent of products, which account for only 20 percent of sales, pulling production in-synch with actual sales.

Calculating Min/Max Levels

One final question is where to set your Min and Max levels. The following general guidelines, which refer back to Chapter 6 on statistics, are recommended for determining both Min and Max. In the following calculations, the statistics of mean and standard deviation are based on the daily sales patterns of a particular product.

The purpose of Min is to guard against stock outs. Therefore, Min must be sufficient to cover the product's safety stock as well as expected demand during transportation lead time. In other words, Min must be high enough to ensure that the company maintains the desired customer service levels while the product is in transit. Since manufacturing produces product every day, rather than erratically based on order point breaks, the actual manufacturing lead time should no longer contain any order queue time. As a general rule, it is best to reduce manufacturing lead time to the absolute minimum, considering only actual manufacturing processes. Most products managed in this manner will have a manufacturing lead time of 1 day. Therefore, safety stock (SS) will be lower than that required in a typical order point model and can be calculated using the formula below, which is based on the statistical principles discussed earlier in this book:

SS = Standard Deviation of Daily Usage x Square Root of Total LT x Z score

(Note that the Total LT includes the reduced manufacturing lead time plus transportation).

Demand through transportation lead time (DTLT) is simply the average daily demand for the product multiplied by the transportation lead time needed to move the product from the production plant to the warehouse.

The purpose of Max is to ensure that the company achieves production stability by providing a buffer level of inventory above Min. The value of Max depends on the level of stability desired, which you calculate using the bell curve assumption depicted in Chapter 6. For example, if 95 percent stability is desired (i.e., 95 percent of the time inventory will be between the Min and the Max), the bell curve tells you that you must cover +/-2 standard deviations from the mean, or average. As a simplifying assumption, on average, inventory will be half way between Min and Max. Therefore, to achieve 95 percent stability, Min will be 2 standard deviations below the average, and Max will be 2 standard deviations above the average. Therefore, Max will be 4 standard deviations above Min. Figure 11-4 includes the final formulas. Although these formulas are statistically valid, they sometimes drive inventory higher than necessary to achieve the desired service and stability levels. Often, one or two peak historical data points can result in an inflated standard deviation, driving this too-high inventory result. Eliminating these points from the data, either by mathematical statistical-outlier logic, or a management decision to simply remove one or two peak data points, can have a dramatic affect on the resulting Min and Max levels. Closely monitoring both service and inventory metrics will tell you if you've cut inventory too deeply, or if there is additional opportunity to reduce inventory further. (This point is equally valid to our earlier safety stock discussion and statistical calculations.)

SCM Team Exercise: Create a new ABC analysis on your stocked finished goods. (Remember that the ABC exercise you have used up to now has been on plant inventory; raw materials, components, etc. Refer to Chapter 3 for a refresher on setting ABC codes). Choose a couple of your highest-volume products from your new ABC analysis, then determine their Min & Max levels using the formulas provided in this chapter. As with the raw material ABC exercise, you may need to work with your IT people to obtain the daily usage data necessary to perform the statistical analysis. Discuss as a team what level of safety stock (sigma level for Min) and production stability (sigma level between Min and Max) makes sense for your company. Be prepared to defend these numbers to your management team.

Look at the new ABC analysis. How many finished goods part numbers make up the top 80 percent of volume? What do these products have in common (shared production lines, common materials, common suppliers, common customers, other)? What potential obstacles do you see in implementing the Min/Max stabilization process described in this chapter (long set up times, disabling current scheduling processes, organizational resistance, etc.)? How can you minimize or overcome these obstacles? Remember that stabilizing production of these products will greatly improve the overall effectiveness of your entire shopfloor and supply chain kanban processes. Keep in mind the lean adage: Replace "no because" with "yes if." (Replace "no, we can't do that because this obstacle exists" with "yes, we can do that if we eliminate this obstacle by doing . . .").

CONCLUSION

The SCM team now has an understanding of the fundamentals of kanban pull processes that improve material flow and supply chain management and can *put kanban into practice*. In other words, you now have the roadmap for putting the fundamental kanban concepts to work in a way that is *immediately actionable and immediately enhances supply chain management*. In addition, you should now be well versed in the terminology associated with kanban and can communicate the principles and practices of kanban to everyone involved in your transformation to kanban. As this workbook has emphasized throughout, a key success factor in implementing and maintaining kanban is clear, consistent communication, with all parties involved in material replenishment abiding by the same principles and understanding the same terminology or using a common language. So where do you go from here?

What You Have Learned—What You Need to Practice

The following are some of the concepts and activities you have already begun to implement through the team exercises. Remember to practice, develop, and continuously improve each of these as you move forward.

- Understand how JIT uses the kanban system to pull materials into the plant and through the operation. Avoid excess inventory by buying or producing in sync with real customer demand and avoid shortages through the proper assignment of safety stock.
- Partner with suppliers to reduce replenishment lead times to enhance just-in-time performance and reduce the amount of inventory required to sustain the manufacturing process.
- Develop greater simplicity using manual-visual inventory management and control processes.
- Apply standard formulas and statistics to determine order points, safety stock, demand through lead time, and lot sizes. Determine ABC class codes and economic order quantity (EOQ). Apply standard deviation and Z-scores in statistical analyses.
- Order parts at the right time based on actual customer usage (JIT).
- Depend less on "just-in-case" inventory and reduce excess inventory (waste).
- Lay out parts so they are easy to find and are located at the point of use.

- Eliminate the need for kitting (non-value-add activity). Replace with kanban move cards to pull parts into manufacturing as needed.
- Improve parts availability.
- Improve manufacturing productivity.
- Improve supplier relationships and communication.

- Understanding how JIT uses kanban
- Partnering with suppliers
- Kanban's simple manual-visual controls
- Applying standard formulas and statistics to determine order point, demand through lead time, lot size, safety stock, ABC codes, and EOQ
- Ordering parts at the right time based on actual customer usage
- Depending less on just-in-case inventory and reducing excess inventory (waste)
- Laying out parts that are easy to find and are located at point of use
- Eliminate kitting by using kanban move cards

Conclusion Figure 1. Kanban Benefits: What You Have Learned— What You Need to Practice

Next Steps

You are now ready to begin the kanban implementation project. Team members understand the concepts and how to make them happen. Management has been brought along through discussions included in team exercises along the way. The team has debated strategies for bringing suppliers on board and for dealing with issues that may arise during the implementation and beyond. It is now time for the team, supported by management, to turn its attention to preparing the organization for the conversion to kanban.

- **Education on lean, JIT:** Use the materials in this workbook to educate the workforce on the changes to come. Paint a picture of what the future state will look like and the project plan you intend to follow to get there.
- **Exercises:** Revisit each of the exercises in this workbook and decide which are appropriate to include in detailed training for various individuals and departments in your organization.

- **Training:** Decide whom to train in your organization. For example, material planners will have to be trained in the details of kanban analysis, manufacturing engineers will need to learn kanban shopfloor layout concepts, and material handlers will need to understand the specific kanban processes associated with pulling kanban cards and managing single- and multi-bin kanbans. Refer to your Gap Analysis from Chapter 10 to guide you in planning appropriate training.

- **Management:** Refer all members of the management staff to the Introduction and Chapters 1, 2, and 10 of this workbook. Form a management steering committee to help guide the project and to provide a ready forum to deal with organizational or other obstacles you encounter along the way. Schedule regular steering committee meetings; weekly during the high periods of activity and less frequently during periods where the team and management agree that less management attention is needed.

- **Financials:** Work with the finance department to develop a means to capture and share the financial benefits of kanban. Clearly establish how much inventory was carried on average for a representative period before implementing kanban, the relationship between inventory and sales (inventory turns), customer service levels, average plant productivity, quality metrics, and any other indicators that management believes are key to measuring performance. Monitor the changes in these metrics as the project moves forward.

- **Developing suppliers:** Establish a communication plan for suppliers, especially those you consider key to your success. Inform them of your plans to implement kanban and discuss potential impact the change may have on them. Invite key suppliers to participate directly in your implementation project. This will serve to educate them in the processes (which may inspire them to do the same in their operations) and to identify and resolve potential supplier issues quickly.

- **Pilot.** Finally, revisit your project plan with the steering committee and begin your pilot.

Final Word on Lean

As stated in Chapter 1 (Figure 1-4), the key goals of lean are *defining a value stream, creating flow, pulling product through the process,* and *promoting continuous improvement.* The kanban fundamentals presented in this workbook touch on each of these elements. The level-loading technique described in Chapter 11, combined with balancing lines to takt time, work together to create smooth repeatable flow through the entire

- **Education**—Educate and train the workforce on what is coming.

- **Management**—Include management in education. Form a management steering committee.

- **Finance**—Capture representative "before kanban" inventory levels. Create metrics to track improvements.

- **Supplier development**—Educate suppliers on kanban and work with them on each of the key touch points discussed in this workbook.

- **Pilot**—Go do it!

Conclusion Figure 2. Kanban Implementation: Next Steps

supply chain (from customers back to manufacturing and upstream to suppliers), with kanban being the tool for *pulling* product through each link in the chain. Continuous improvement is achieved by rigorously following the kanban analysis and exception reporting described throughout this workbook and by ensuring these activities meet their intended aims through routine kanban audits.

It is my hope that you can now clearly see the benefits of kanban for managing the supply chain in the world of *lean* and that you can immediately apply the principles and techniques of kanban to your operation. When properly implemented, kanban will improve your material flow, thereby increasing the productivity, profitability, and performance of the supply chains so critical to sustaining lean plant operations. In today's marketplace of global competition, outsourcing, and lean thinking, the very survival of your company may depend on implementing the simple but effective rules of kanban at the heart of *your* lean, just-in-time enterprise.

INDEX

125

ABOUT THE AUTHOR

Steve Cimorelli is the Director of Inventory Control and Forecasting with Fleetguard, Inc., a global manufacturer of diesel engine filtration and exhaust products. He holds a BS degree in Industrial Engineering from the University of Central Florida and is certified at the Fellow level by APICS in production and inventory management (CFPIM). He has over 25 years experience in manufacturing, engineering, materials, and computer systems in a variety of industries, including aerospace and defense, industrial equipment and commercial products.

From 2001 to 2004, Steve consulted with numerous manufacturing plants in the United States, Mexico, Canada, and Europe, developing and implementing lean manufacturing and materials excellence practices, especially in the areas of kanban, forecasting, line balancing, and production stabilization.

Steve was first introduced to lean manufacturing and just-in-time principles in the late 1980s while working towards his certification in production and inventory management (CPIM) with APICS, The Association for Operations Management. As he explains, "These ideas resonated so strongly with me that I began to actively seek opportunities to put them into practice. It wasn't until 1995, when I joined Square D–Schneider Electric that I saw first-hand how they actually worked. From that point on, I became a vocal advocate and implementer of these practices throughout Square D and outside of the company through consulting work, teaching, and publishing."

Steve's earlier published works include a chapter titled "Control of Production and Materials" in the *Handbook of Manufacturing Engineering* (published in 1995 by Marcel Dekker) and the February 2002 cover story of "APICS—The Performance Advantage" magazine titled "Reduced Demand." He is an active member of the APICS Space Coast Chapter and a frequent speaker at APICS sponsored events.

Steve lives in Titusville, Florida with his wife, Cindi, and faithful dog, Ranger. His grown children are students in the Florida state university system.